American Institute of Chemical Engineers

Conference on Engineering Sciences and Fundamentals

Presentations at the
2008 AIChE Spring National Meeting

April 6-10, 2008
New Orleans, Louisiana, USA

Printed from e-media with permission by:

Curran Associates, Inc.
57 Morehouse Lane
Red Hook, NY 12571
www.proceedings.com

ISBN: 978-1-60560-217-2

ISBN: 978-1-60560-217-2

American Institute of Chemical Engineers
Proceedings
Three Park Avenue
New York, NY 10016-5991
Phone: 212-591-8100

www.aiche.org

American Institute of Chemical Engineers

Conference on Engineering Sciences and Fundamentals
2008

TABLE OF CONTENTS

A Group Contribution Based-Model to Estimate Free Volume in Amorphous Polymers for Solvent Diffusion

Hong-ling Lv, Bao-guo Wang, Ji-chu Yang*

Department of Chemical Engineering, Tsinghua University, Beijing 100084, the People's Republic of China

**corresponding author: bgwang@tsinghua.edu.cn*

Introduction

The transport of organic solvents through polymers plays a crucial role in many different types of polymer-related technological and industrial applications, such as polymerization and membrane separation. Nowadays, the widely accepted approach to describe transport properties in polymer solutions is the free-volume theory, which emphasizes on the amount of "empty space" as the dominant factor for diffusion. The model proposed by Vrentas and Duda is a representative of this theory, in which most parameters can be obtained from pure component data and no adjustable parameters are used [1-2]. However, free volume parameters of polymer are usually determined by fitting the results from measurement of polymer viscoelasticity, meaning a great deal of time and cost consumption. In order to overcome this shortcoming, our previous study has employed equations-of-state (EOSs) to estimate polymer free volume [3, 4], and obtained good consistence of calculations with experiments. Even though, the characteristic parameters of polymers in the EOSs, which are available in the literature, are limited only to some commonly-used polymers.

In this paper, a group contribution-based method will be presented to predict solvent diffusion coefficients with both infinite dilute and normal solvent concentration in polymers, by estimating polymer free volume with an EOS. The published positron annihilation lifetime spectroscopy (PALS) results will also be employed to evaluate the modification of the diffusion model.

Theory

The Vrentas-Duda model

According to free-volume theory [1-2], the specific volume of a liquid is considered as the sum of three parts: the core volume, the interstitial free volume and the hole free volume. Only the third part is effective for molecular transport. In a mixture of a simple organic solvent and a polymeric liquid, the total free volume available for transport is contributed by both the solvent and polymer. In particular, for a polymer-solvent system with vanishing solvent concentration, the infinite dilution diffusion coefficient, D^{∞}, can be expressed as follows

$$D^{\infty} = D_0 \exp\left(-\frac{E^*}{RT}\right)\exp\left(-\frac{\xi V_2^*}{V_{\mathrm{f}}}\right) \tag{1}$$

where D_0 is a constant pre-exponential factor, E^* is the energy which a molecule must possess to overcome attractive forces holding it to its neighbors, V_2^* is the specific hole free volume of polymer required for molecular jump, V_f is the hole free volume provided by polymer for solvent diffusion, and ξ is the ratio of the critical molar volume of the solvent jumping unit to that of the polymer jumping unit, which can be estimated by an empirical relationship [5].

The GCLF-EOS model

The group-contribution lattice-fluid equation-of-state (GCLF-EOS) model is based on the lattice theory [6], by introducing a non-random term related to the local composition effects for molecules, and with the assumption that the holes in the lattice are randomly distributed. The GCLF-EOS has been successfully used in the prediction of polymer density, phase equilibriums of solvent and polymer, solubility parameters and glass transition temperatures. A significant advantage of this model is that only the structures of the molecules in the systems are required in the calculation. The expression of the GCLF-EOS and the group parameters for 37 common functional groups are presented in the literature [6].

The modified free-volume model

Following the previous consideration [3, 4, 7], the hole free volume of polymer can be interpreted as the free volume in the Williams-Landel-Ferry (WLF) theory:

$$V_f = V(T)\left[0.025 + \ln\frac{V(T)}{V(T_{g2})}\right] \tag{2}$$

Here, V is the specific volume of polymer, which can be determined by the GCLF-EOS, and T_{g2} is the glass transition temperature of polymer.

The modified version of the free-volume model can be expressed by combining eq. (1) and eq. (2), given as follows

$$D^\infty = D_0 \exp\left(-\frac{E^*}{RT}\right)\exp\left\{-\frac{\xi V_2^*}{V(T)\left[0.025 + \ln\left(V(T)/V(T_{g2})\right)\right]}\right\} \tag{3}$$

The GCLF-EOS is a purely predictive theory, requiring only the structure of the molecules. Therefore, all the parameters with relevance to polymer in this model can be determined by group contribution method, and the complicated process of measuring polymer viscoelasticity can be avoided.

Results and Discussion

Calculation of solvent diffusion coefficients

The infinite dilution diffusion coefficient provides a useful way to evaluate the proposed method by estimating free volume in polymers, since the free volume available for solvent diffusion is only contributed by the polymer. As can be seen in Figs. 1 and 2, the predictions of infinite dilution diffusion coefficients of ethylbenzene and heptane in poly(vinyl acetate) are generally consistent with the experimental data, and the group contribution-based approach is valid to estimate polymer free volume for solvent diffusions.

The self-diffusion accounts for the molecular behaviors in terms of the Brownian motion, in which the influence of concentration gradient can be reasonably ignored. Fig. 3 illustrates the calculations of cyclohexane self-diffusion coefficients in polybutadiene with both the modified and Vrentas-Duda models. The modified model provides better accuracy than the original model, due to the correct estimation of the polymer free volume with the GCLF-EOS. Moreover, this model requires no measurements of polymer viscoelasticity.

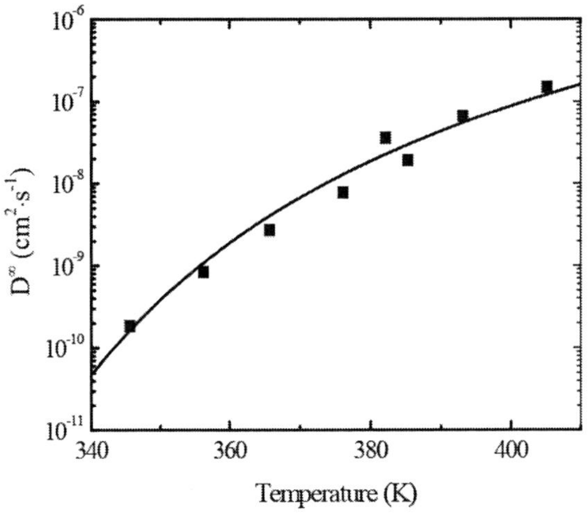

Fig.1 Experimental data and calculation of infinite dilution diffusion coefficients of ethylbenzene in poly(vinyl acetate)

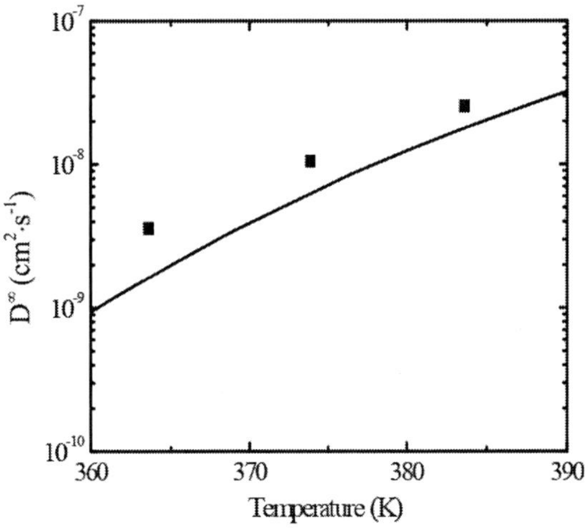

Fig.2 Experimental data and calculation of infinite dilution diffusion coefficients of heptane in poly(vinyl acetate)

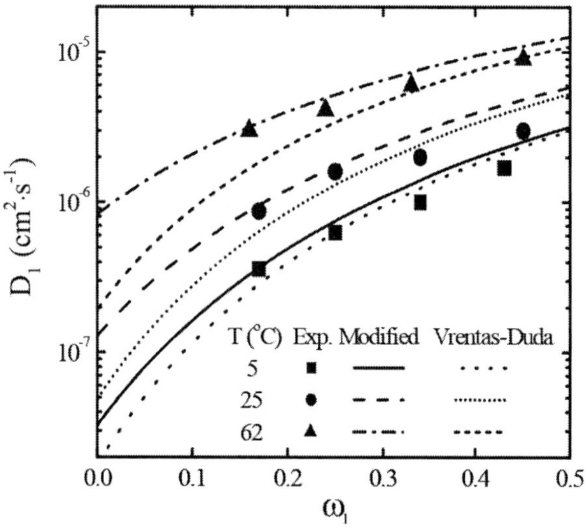

Fig.3 Comparison of cyclohexane self-diffusion coefficients in polybutadiene with the calculation from the modified and Vrentas-Duda models

Analysis from PALS results

PALS is a unique technique to investigate free-volume holes in polymers, which is based on the principle that positronium (Ps) atoms are formed and preferentially localized in holes. When the effect of energy on molecular motion is neglected within a limited temperature range, only the portion with enough size rather than all the holes in polymers can be used for molecular motion. The minimum or critical hole size should have the similar meaning with the size determined by the parameter V_2^* in Eq.(1). The ratio of the free volume available for solvent diffusion to the total free volume in polymers can be estimated by the Simha-Somcynsky EOS: $V_f/\left[(1-y)V\right]$. For polystyrene at 463 K, this value is estimated to be 75.5 %. Using this data and the published PALS measurements, the value of critical hole radius is calculated to be ~3.21 Å with the assumption of spherical holes. This value is comparable to the value determined from V_2^* (3.27 Å), meaning that free volume of polymers estimated by the group contribution method can reflect the actual situation for molecular diffusion.

Conclusions

A group contribution-based approach has been established to estimate solvent diffusion coefficients, and then the free volume of polymers becomes predictive without knowledge of any diffusion or viscoelastic data. Predictions of solvent diffusion coefficient were generally in good agreement with experimental data. The PALS results have also been investigated to show the reliability of the proposed method.

Acknowledgement

The authors gratefully acknowledge the financial assistance from the Major State Basic Research Development Program of China (973 Program) (No. 2003CB615701) and the National Natural Science Foundation of China (No. 20676068).

References

1. Vrentas JS, Duda JL. Diffusion in polymer-solvent systems. I. Reexamination of the free-volume theory. J Polym Sci Polym Ed, 1977, 15, 403-416(a) ; 417-439.(b).
2. Hong SU. Prediction of polymer/solvent diffusion behavior using free-volume theory. Ind Eng Chem Res, 1995, 34, 2536-2544.
3. Lv HL, Wang BG. Prediction of solvent diffusion coefficient in amorphous polymers based on an equation-of-state. J Polym Sci Polym Phys, 2006, 44, 1000-1009.
4. Wang BG, Lv HL, Yang JC. Estimation of solvent diffusion coefficient in amorphous polymers using the Sanchez-Lacombe equation-of-state. Chem Eng Sci, 2007, 62, 775-782.
5. Hong SU. Predicting ability of free-volume theory for solvent self-diffusion coefficients in rubbers. J Appl Polym Sci, 1996, 61, 833-841.
6. Lee BC, Danner RP. Prediction of polymer-solvent phase equilibria by a modified group-contribution EOS. AIChE J, 1996, 42, 837-849.
7. Lv HL, Wang BG, Yang JC. A group contribution-based model to predict organic solvent diffusivities in amorphous rubbery polymers. Polym J, 2007, 39, 1167-1171.

A New Methodology for the Determination of Homogeneous Diffusion Coefficients of Biomolecules in Adsorbents via Non-Equilibrium Modeling

Gönül Akkaya and Ahmet R. Özdural

Chemical Engineering Department, Hacettepe University, Beytepe, 06800 Ankara, Turkey,

Abstract

The present work describes a new technique for the estimation of homogeneous solid diffusivity, D_s for resins through batch uptake experiments. The model is developed for the nonlinear adsorption isotherm case, and it takes non-equilibrium distribution between the bulk liquid and the solid concentration into account. We modified the conventional batch adsorption analysis by introducing an expression that enables the prediction of solid and liquid concentrations at the adsorbent particle-liquid interphase. During the experimental phase of this study batch uptake experiments were carried out with α-amylase, BSA and IgG, with molecular masses ranging from 51 000 to 160 000, onto a strong anion exchanger Q Sepharose XL. For these solutes and the adsorbent media, single component homogeneous diffusion coefficients were experimentally determined with the non-equilibrium methodology, introduced in the present study.

Keywords: Homogeneous diffusion; Transport properties; Mass transfer effects; Stationary phases; Non-equilibrium adsorption.

1. Introduction

When characterizing the adsorption phenomena, one is faced with the choice between the assumptions of the local equilibrium conditions (Coppola and Levan, 1983; Frey et al., 2002) or non-equilibrium conditions (Filippov, 1994; Schork and Fair, 1998). In packed-bed columns, adsorption is generally assumed to be at equilibrium locally, i.e. instant equilibrium is assumed between the solid and the fluid phases. Although the principle of local equilibrium has been traditionally employed for a wide variety of adsorption studies, it was shown that models based on assumptions of local equilibrium cannot quantitatively describe behavior of a typical industrial adsorption system (Harwell et al., 1980). During packed bed adsorption, Ruthven (1984) indicated that the initial breakthrough time is more sensitive to mass transfer resistance than to axial dispersion whereas the situation is reversed in the saturation region. For liquid chromatography simulation studies, equilibrium dispersive model (Guiochon et al., 1994) is employed by many authors. The equilibrium-dispersive assumption, which neglects all transient resistances, makes it impracticable to differentiate the effects of external and internal mass transfer resistances. In ideal nonlinear chromatography, in order to account for zone broadening, an apparent axial dispersion term is used to describe and understand the band broadening effects of both axial dispersion and the finite rate of the mass transfer kinetics (Guiochon, 2002). For small molecules and highly efficient columns this model is generally adequate. However, especially in the case of large molecules, the assumption of equilibrium conditions is not valid in the dynamic state that exists in a chromatographic column (Skoog et al., 2003). Özdural et al. (2004) showed that local equilibrium based frontal chromatographic column simulations become relatively free from errors only at near saturation conditions.

The simple homogeneous diffusion model appears to predict the experimentally observed trends with respect to protein concentration and boundary layer mass transfer effects for the gel-composite adsorbents (Weaver and Carta, 1996; Hunter and Carta, 2002). In this model, all solute within the particle, whether in the pores or adsorbed by the solid network phase of the particle is lumped into a single quantity. It is assumed that intraparticle mass transfer occurs with a driving force governed by the lumped solute concentration within the particle. Other diffusion models, such as the pore diffusion and the surface diffusion, are also employed by many authors (Skidmore et al., 1990; Yoshida et al., 1994; Tonga et al., 1994; Weaver and Carta, 1996; Chang and Lenhoff, 1998; Hunter and Carta, 2000; Chen et al., 2002; Tscheliessnig et al., 2005; Gao et al., 2006) as well.

Several published studies dealt with the determination of diffusivity proteins in solids through protein uptake experiments in a stirred batch apparatus (Weaver and Carta, 1996; Hunter and Carta; 2000; Quinones-Garcia et al., 2001; Carta and Ubiera, 2003; Tscheliessnig et al., 2005; Gao et al., 2006; Zhou et al., 2006) where instant equilibrium is assumed between the solid and the liquid phases. On the other hand, the assessment presented above reveals that in order to reach a factual batch uptake adsorption dynamics portrait the inclusion of non-equilibrium constraints in the modeling studies is essential. This is especially significant for the case of larger, slowly diffusing molecules, such as proteins. It might be interestingly enough to state that there seems to be extremely little published information regarding the non-equilibrium adsorption and transport properties that govern batch uptake kinetics.

In the present study homogeneous diffusion model is assumed with respect to protein adsorption. Our main purpose is the introduction of non-equilibrium approach, originally proposed for chromatographic columns modeling (Özdural et al., 2004), to the determination of diffusivity proteins in adsorbents through protein uptake experiments in a stirred batch apparatus. The significance of this work is twofold. First, it delineates the importance of the non-equilibrium conditions that prevails throughout the batch uptake experiments. Second, shows the relationship of dynamic behavior of interphase and bulk liquid concentrations.

2. Theoretical

For the model development, a batch uptake system is envisaged where single component adsorption takes place. Non-equilibrium conditions exist between the spherical adsorbent particle and the liquid in the perfectly mixed reservoir. It is assumed that the non-linear equilibrium data can be represented by Langmuir equation. The model is based on a dual resistance model combining external mass transfer and intraparticle transport by homogeneous diffusion, and assumes a time dependent parabolic concentration profile within the particle.

The solute mass balance around the reservoir shown in Fig. 1 gives:

$$V\frac{dc(t)}{dt} = -\frac{m}{\rho_p}\frac{d\bar{q}(t)}{dt} \qquad (1)$$

where $c(t)$ is the time dependent bulk liquid concentration (mg/cm^3), m is the adsorbent weight within the reservoir (g), V is the liquid volume in batch adsorber (cm^3), $\bar{q}(t)$ is the time dependent solid average concentration (mg/cm$^3_{solid}$), ρ_p is the adsorbent particle density (g/cm^3) and t is the time (s). With referring to Fig. 2, the solid uptake rate for spherical adsorbent particles can be expressed as following (Radcliffe et al., 1982; Cooney, 1991).

$$\frac{d\bar{q}(t)}{dt} = k_f \frac{3}{r_p}\left[c(t) - c_s^*(t)\right] \tag{2}$$

where $c_s^*(t)$ is the time dependent interphase liquid concentration (mg/cm^3), r_p is the particle radius (cm) and k_f is the liquid film mass transfer coefficient (cm/s). Substitution of Eq. (2) into Eq. (1) gives the following equation (Hahn et al., 2005).

$$V\frac{dc(t)}{dt} = -\frac{m}{\rho_p} k_f \frac{3}{r_p}\left[c(t) - c_s^*(t)\right] \tag{3}$$

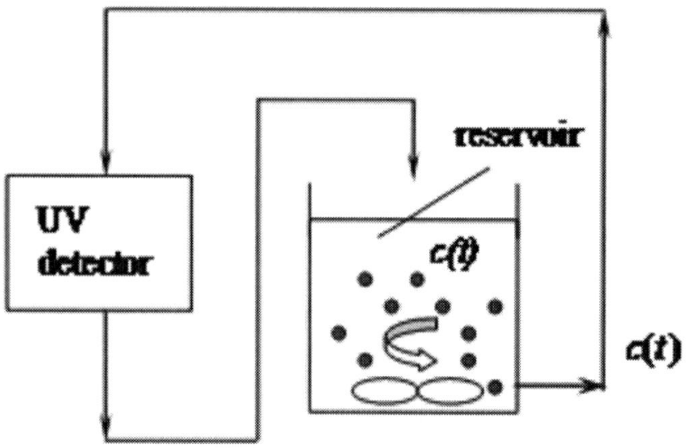

Figure 1. Schematic view of the stirred batch uptake system.

2.1. Estimation of film mass transfer coefficient in batch systems

It is generally agreed that, for favorable isotherms, during the early stages of batch uptake from dilute solutions the value of $c_s^*(t)$ becomes negligible (Mc Cabe et al., 1985). Under these conditions integrating Eq. (3) between the limits of $t = 0$, $c = c_0$; $t = t$, $c = c(t)$ gives:

$$\ln\frac{c}{c_0} = -\frac{3m}{\rho_p r_p V} k_f t \tag{4}$$

Eq. (4) indicates that k_f value can easily be determined for short times and low concentration batch experiment uptake data as long as the adsorption isotherm is favorable.

2.2. Estimation of homogeneous diffusivity in batch systems

Once film mass transfer coefficient value is obtained from a low concentration experimental data, then homogeneous diffusivity, D_s (cm^2/s) can be calculated from a high concentration experiment, by keeping all other experimental parameters the same with that of the low concentration experimentation. It is clear that for high concentration experiments the $c_s^* \approx 0$ assumption is no longer valid. Thus, in order to predict the change of bulk liquid concentration with time, one should incorporate c_s^* to the solution of Eq. (3). In this work, the widely used immediate equilibrium assumption between the solid average concentration and the bulk liquid concentration is ruled out due to the fact that under dynamic conditions with realistic mass transfer resistances this assumption is not valid. However it is reasonable to assume that immediate equilibrium is attained at the interface since transfer of solute at the interphase to adsorbed state is generally very fast. Fig. 2 visualizes the liquid and solid concentrations in homogeneous adsorption model. Eq. (5) gives well known Langmuir adsorption isotherm expression expressed in terms of interphase concentrations.

$$q^*_s = \frac{q_m K_L c_s^*}{1 + K_L c_s^*} \tag{5}$$

The q_m (mg/cm^3_{solid} and K_L (cm^3/mg) terms in Eq. (5) are the Langmuir constants and q^*_s is the solid surface concentration in equilibrium with c_s^*. As long as Langmuir adsorption isotherm holds, adaptation of Özdural et al. (2004) methodology to the batch systems shows that interphase liquid concentration, c^*_s can be given by Eq. (6). It is likely to come across with an implicit equation for the evaluation of c^*_s for other non-linear adsorption isotherms.

$$c_s^* = \frac{-M + \sqrt{M^2 + \frac{4}{K_L}\left[c + \frac{5\bar{q}}{Bi}\right]}}{2} \tag{6}$$

$$M = \frac{5q_m}{Bi} - \frac{5\bar{q}}{Bi} + \frac{1}{K_L} - c \tag{7}$$

$$Bi = k_f r_p / D_s \tag{8}$$

Where Bi is the Biot number. Regarding the mass transfer resistances, Eq. (3) contains the k_f term only, but as shown in Eqs. (6) to (8) solid internal resistance is inherently included in the c_s^* term.

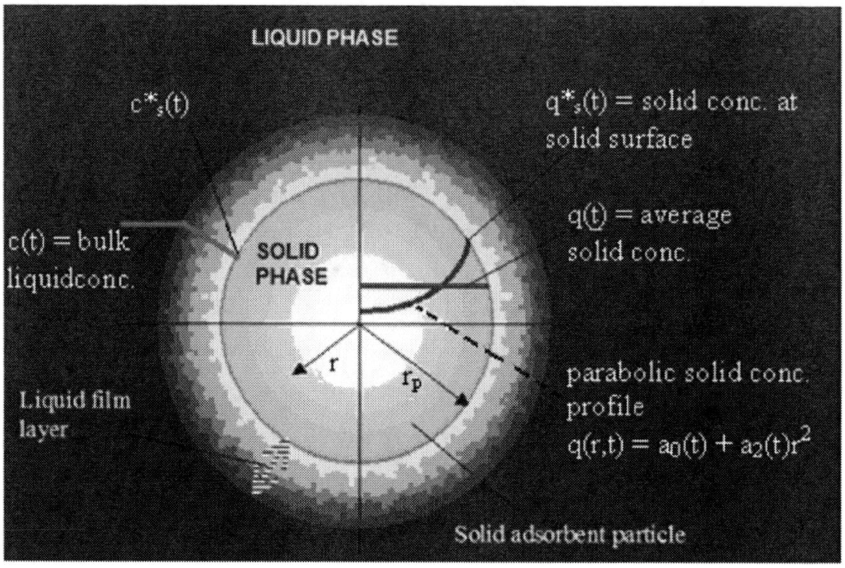

Figure 2. Visualization of liquid and solid concentrations in homogeneous adsorption model.

Another feature of the non-equilibrium model presented in this study is that it obviates the usually employed solution of coupled ordinary and partial differential equation systems (McCue et al., 2003) and makes it possible to use the well known Runge-Kutta algorithms during simulation studies. Eq. (3) can now be solved without assuming $c_s^* \approx 0$ since Eq. (6) gives c_s^* value. During the numerical solution procedure, the $\bar{q}$ value at j time panel is evaluated from the c and $\bar{q}$ values at j-1 time panel through the mass balance methodology of Özdural et al. (2004).

$$\bar{q}_j = \bar{q}_{j-1} + \frac{V}{\dfrac{m}{\rho_p}}(c_j - c_{j-1}) \tag{9}$$

For finding homogeneous diffusivity, model predicted reservoir concentration versus time data for different D_s values were numerically generated and compared with the experimentally gathered reservoir concentration versus time data until the two profiles agree. D_s value which gives the best fit, evaluated through the root-mean-square (rms) analysis, was taken as the homogeneous solid diffusion coefficient of the system of interest.

3. Experimental

The experiments presented were conducted with α-amylase, BSA and IgG, onto a strong anion exchanger Q Sepharose XL, which is an industrially relevant ion exchange adsorbent specially designed for use in packed beds to capture biomolecules from clarified feed-stocks.

4. Results and Discussion

The equilibrium uptake isotherms for IgG, BSA and a-amylase on Q Sepharose XL were nonlinear and Langmuir adsorption isotherm parameters, q_m and K_L along with experimentally determined homogeneous diffusion coefficient values are summarized in Table 1.

Table 1. Equilibrium parameters and homogeneous diffusivities

Protein	M.W. (kDa)	q_m (mg/cm^3)	K_L (cm^3/mg)	$D_s \times 10^9$ (cm^2/s)
α-amylase	51	161.3	12.4	1.38
BSA	66	153.9	32.5	1.05
IgG	160	170.1	11.3	0.92

For the present experimental conditions k_f values, calculated through Eq. (4), were 0.78×10^{-3} cm/s, 2.92 x !0^{-3} cm/s and 0.19×10^{-3} cm/s for α-amylase, BSA and IgG respectively. Fig. 3 illustrates the experimental concentration profiles and model outputs obtained for α-amylase, as an example.

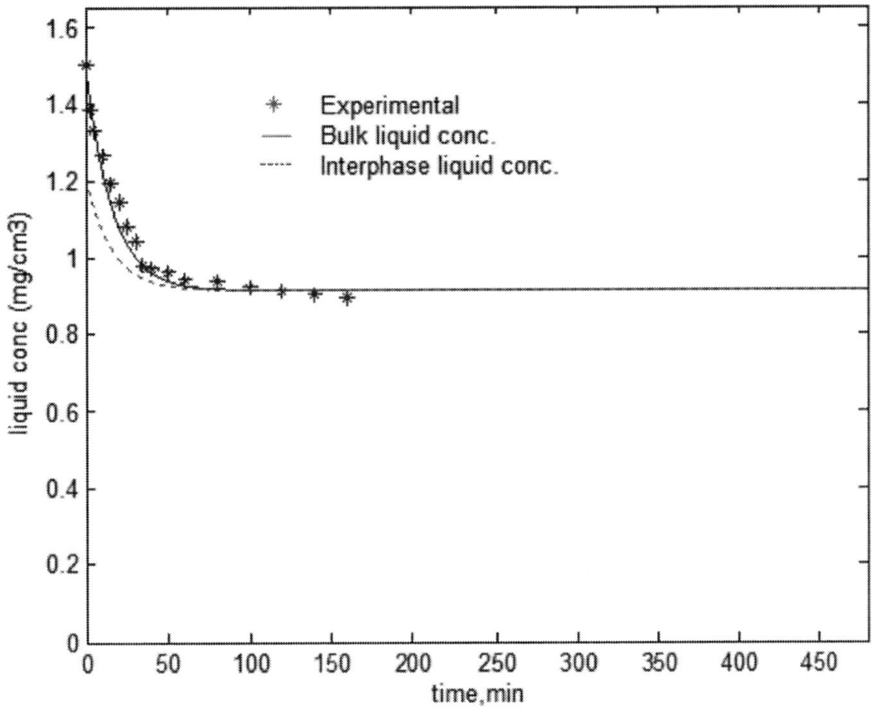

Figure 3. Comparison of the experimental concentration profiles with model outputs obtained for α-amylase.

This figure illustrates that, the difference between interphase and bulk liquid concentrations are evident. It was found that the homogeneous diffusion model successfully predicts the observed uptake trends. It was found that D_s values decreases as the protein M.W. increases. Furthermore, it was concluded that the non-equilibrium methodology introduced here might become a standard tool for the determination of homogeneous diffusion coefficients of biomolecules on adsorbents.

References

Carta, G., and Ubiera, A., Particle-size distribution effects in batch adsorption, *AIChE J.*, 49 (2003) 3066.

Chang, C., Lenhoff, A.M., Comparison of protein adsorption isotherms and uptake rates in preparative cation-exchange materials, *J. Chromatogr. A*, 827 (1998) 281.

Chen, W.D., Dong, X.Y., Sun, Y., Analysis of diffusion models for protein adsorption to porous anion-exchange adsorbent, *J. Chromatogr. A*, 962 (2002) 29.

Cooney, D.O., Determining external film mass transfer coefficients for adsorption columns, *AIChE J.*, 37 (1991) 1274.

Filippov, L.K., Coherent and incoherent frontal patterns of multicomponent adsorption dynamics for variable linear fluid velocity in the fixed bed - I. Frontal patterns for linear adsorption isotherms, *Chem. Eng. Sci.*, 49 (1994) 3499.

Frey, D.D., Narahari, C.R., Butler, C.D., General local-equilibrium chromatographic theory for eluents containing adsorbing buffers, *AIChE J.*, 48 (2002) 561.

Gao, D., Lin, D-Q., Yao, S-J., Protein adsorption kinetics of mixed-mode adsorbent with benzylamine as functional ligand, *Chem. Eng. Sci.*, 61 (2006) 7260.

Glueckauf, E. and Coates, J.I., Theory of Chromatography. Part IV. The influence of incomplete equilibrium on the front boundary of chromatograms and on effectiveness of separation, Journal of the Chemical Society (1947), 1315.

Guiochon, G., Golshan Shirazi, S., Kati, A.M., Fundamentals of Preparative and Nonlinear Chromatography. Academic Press, Boston, 1994.

Guiochon, G., Preparative liquid chromatography, *J. Chromatogr. A*, 965 (2002) 129.

Hahn, R., Bauerhansl, P., Shimahara, K., Wizniewski, C., Anne Tscheliessnig, A., Jungbauer, A., Comparison of protein A affinity sorbents II. Mass transfer properties, *J. Chromatogr. A*, 1093 (2005) 98.

Harwell, J.H., Liapis, A.I., Litchfield, R., Hanson, D.T., A non-equilibrium model for fixed-bed multi-component adiabatic adsorption, *Chem. Eng. Sci.*, 35 (1980) 2287.

Hunter, A. K., Carta, G., Protein adsorption on novel acrylamido- based polymeric ion exchangers-II. Adsorption rates and column behavior, *J. Chromatogr. A*, 897 (2000) 81.

Hunter, A.K., Carta, G., Protein adsorption on novel acrylamido-based polymeric ion-exchangers IV. Effects of protein size on adsorption capacity and rate, *J. Chromatogr. A*, 971 (2002) 105.

McCabe, W.L., Smith, J.C. Harriot, P., Unit Operations of Chemical Engineering, 4th ed., McGraw-Hill, New York, 1985.

McCue, J.T., Kemp, G., Low, D., Quinones-Garcia, I., Evaluation of protein-A chromatography media, *J. Chromatogr. A*, 989 (2003) 139.

Özdural, A.R., Alkan, A., Kerkhof, P.A.J.M., Modeling chromatographic columns: Non-equilibrium packed-bed adsorption with non-linear adsorption isotherms, *J. Chromatogr. A*, 1041 (2004) 77.

Quinones-Garcia, I., Rayner, I., Levison, P.R., Dickson, N., Purdom, G., Performance comparison of suspended bed and batch contactor chromatography, *J. Chromatogr. A*, 908 (2001) 169.

Radcliffe, D.F., Leng, L., Thomas, G., Direct measurement of external mass-transfer in packed sorbent beds, *AIChE J.*, 28 (1982) 344.

Schork, J.M., Fair, J.R., Parametric analysis of thermal regeneration of adsorption beds, *Ind. Eng. Chem. Res.*, 27 (1998) 457.

Skidmore, G. L., Horstmann, B. J., Chase, H. A., Modeling single component protein adsorption to the cation exchanger S Sepharose FF, *J. Chromatogr.*, 498 (1990) 113.

Skoog, D.A. West, D.M. Holler, F. J. Crouch, S.R. Fundamentals of Analytical Chemistry, 8th ed., Brooks Cole, Belmont, 2003.

Tonga, A., Liapis, A.I., Sieher, D.J., Equilibrium and kinetic parameters of the adsorption of α-Chymotrypsinogen A onto hydrophobic porous adsorbent particles, *J. Chromatogr. A*, 686 (1994) 21.

Tscheliessnig, A., Hahn, R., Jungbauer, A., In situ determination of adsorption kinetics of proteins in a finite bath, *J. Chromatogr. A*, 1069 (2005) 23.

Weaver, L. E., Carta, G., Protein adsorption on cation exchangers: Comparison of macroporous and gel-composite media, *Biotechnol. Progr.*, 12 (1996) 342.

Yao, C., Tien, C., Approximation of intraparticle mass transfer in adsorption processes—I. Linear systems, Chem. Eng. Sci., 47 (1992) 457.

Yoshida, H., Yoshikawa, M., Kataoka, T., Parallel transport of BSA by surface and pore diffusion in strongly basic chitosan, *AIChE J.*, 40 (1994) 2034.

Zhou, X-P., Li, W., Shi, Q-H., Sun, Y., Analysis of mass transport models for protein adsorption to cation exchanger by visualization with confocal laser scanning microscopy, *J. Chromatogr. A*, 1103 (2006) 110.

139c Molecular Modeling Study of Adsorption and Diffusion of Mixed Gases in Cubtc Metal Organic Framework

Karl Johnson[1], Jinchen Liu[1], Seda Keskin[2], and David S. Sholl[2]. (1) University of Pittsburgh, 1249 Benedum Hall, Department of Chemical Engineering, Pittsburgh, PA 15261, (2) Carnegie Mellon University, Department of Chemical Engineering, Carnegie Mellon University, Pittsburgh, PA 15213

We use atomically detailed simulations and continuum theory to study the adsorption and diffusion of mixed gases in the metal organic framework known as CuBTC. We have computed pure fluid and mixture isotherms for CH4, CO2, and H2 in CuBTC from grand canonical Monte Carlo simulations. We compare our mixture isotherms with predictions from the ideal adsorbed solution theory. We have used equilibrium molecular dynamics to compute self and transport diffusivities for the pure fluids. We have predicted the mixed gas diffusion coefficients from SSK theory. We have compared the SSK theory predictions with mixed gas diffusivities computed from equilibrium molecular dynamics. We assess the accuracy of the SSK theory and discuss prediction of macroscopic permeance and selectivity based on predictions from simulations and SSK theory.

Liquid Injection into Fluidized Beds

Paul (Pinghua) Zhao
CPFD Software, LLC, Albuquerque, NM 87111

Peter J. O'Rourke
CFD d'OR Software and Consulting, LLC, Los Alamos, NM 87544

Dale Snider
CPFD Software, LLC, Albuquerque, NM 87111

Introduction

Gas-liquid-solid fluidized beds are used in fluid cokers to convert a heavy hydrocarbon feed into lighter hydrocarbon components, gases, and coke [1,2]. The liquid hydrocarbon feed is introduced as a spray that wets hot solid bed particles. The solid bed particles provide a large surface area upon which rapid vaporization and chemical reactions take place because of high heat transfer rates. In such a process, the liquid to solids contacting and the resulting liquid film formation on particles is fundamental. For a given liquid feeding rate, a more uniform liquid to particle coverage results in a reduction in the film thickness leading to lower likelihood of agglomeration, higher heat transfer rates and increased product yields.

The mechanism by which the liquid spreads over the particles' surfaces is collision between liquid spray droplets and bed particles, and between the wet bed particles themselves. More generally, collisions result in mass, momentum, and energy transfer between particles and cause a variety of associated physical effects. Momentum transfer results in the damping of relative motion between particles, which, in turn, reduces the collision frequency. In addition to spreading the liquid over the particles' surfaces, collisional mass transfer results in mixing of liquid residing on different particles and a tendency toward local uniformity of liquid chemical composition and liquid temperature on different particles.

A unified model and simulations for collisional mass, momentum, and energy transfer between bed particles in a gas-liquid-solid fluidized bed is described by O'Rourke, et al [3]. The mathematical model extends the equations of the Multiphase Particle-in-Cell (MP-PIC) [4,5,6] by including collision terms on the right-hand side of the transport equation for the single-particle distribution function. The form of the collision terms is the same as in the BGK model for collisions in the Boltzmann equation of gas dynamics [7]. In the BGK model, the effect of collisions is represented by a simple relaxation term on the right-hand side of the Boltzmann equation that is the difference between the current value of the molecular distribution function and the equilibrium Boltzmann velocity distribution, divided by a molecular collision time.

The situation in gas-liquid-solid fluidized beds is more complicated than in gas dynamics because, in addition to collisions causing particle velocities to relax to an equilibrium value, particle collisions cause transfer and mixing of liquid-film mass and, consequently, result in equilibration of liquid-film masses, chemical compositions, and film temperatures. To represent this situation, we assume that there are two BGK-like terms on the right-hand side of the particle distribution function, with different equilibrium distributions and different collisional time scales. One time scale is associated with achieving particle velocity equilibrium with other particle properties frozen, and the other time scale is associated with achieving full equilibration of film masses, velocities, liquid-film compositions, and temperatures.

In this work, we assume that at equilibrium the ratio of the liquid to solid mass of individual wet particles is a constant equal to the local ratio of the total liquid to total solid mass density. It can easily be seen that this assumption is equivalent to the assumption that at equilibrium, the ratio of the liquid film thickness to solid particle radius is locally a constant and independent of solids radius. Our assumption contrasts with, and improves upon, the assumption made by some authors of a constant film thickness that is independent of the radius of the solid particle to which the film adheres [1].

In our numerical method, integration of the collision terms is performed within the framework of the computational particle fluid dynamic (CPFD®) software in the commercial code Barracuda™ where particle calculations are based on the MP-PIC method [3,4,5]. The continuous distribution function of particle properties, which is obtained by a turbulence averaging procedure, is numerically approximated by a discrete distribution of Lagrangian computational parcels, each of which represents a number of physical particles with identical properties. In this study, we take the properties of physical particles to be their location, solid mass, the masses of the chemical species in their liquid film, and the particle velocity. In this paper, we ignore the solid and liquid film temperatures for brevity; however the inclusion of temperature in the model is similar and straightforward [3].

Particle distribution function equations with collisions

Following the MP-PIC method, the dynamics of the particle field is described by the particle distribution function $f(u_j, x_j, M_s, M_{f,i}, t)$. where x_j is the particle position, u_j is the particle velocity, M_s is the solid mass, $M_{f,i}$ is the liquid film mass for each liquid species i, and t is time. The system is isothermal and the solid mass is constant for each solid (does not fracture). The probability function integrated over all velocities and masses gives the average number of particles per unit volume at x_i and time t. The transport equation for the particle distribution function f is derived in an analogous fashion to the derivation of the Boltzmann equation of gas dynamics [7]:

$$\frac{\partial f}{\partial t}+\frac{\partial(f\dot{x}_j)}{\partial x_j}+\frac{\partial(f\dot{u}_j)}{\partial v_j}+\frac{\partial(f\dot{M}_s)}{\partial M_s}+\frac{\partial(f\dot{M}_{f,i})}{\partial M_{f,i}}=\frac{f^{eq,u}-f}{\tau^u}+\frac{f^{eq,t}-f}{\tau^t} \tag{1}$$

The equilibrium distribution $f^{eq,u}$ is the distribution that results when particle velocities are equilibrated, and the remainder of the particle distribution remains unchanged. The full equilibrium distribution function $f^{eq,t}$ is the particle distribution that results when full collisional equilibrium of all particle quantities is achieved. The two collision times are defined as being proportional to the time between particle collisions τ_{coll}, which is a function of local fluidization conditions including velocity fluctuations, particle size and solid-phase volume fraction:

$$\tau^u=\tau_{coll}/K_u \quad \text{and} \quad \tau^t=\tau_{coll}/K_t \tag{2}$$

where K_u and K_t are positive dimensionless factors that may be functions of local mean particle properties. In this study, we take K_u and K_t to be constants. Physically, K_u and K_t specify the efficiency of collisions in bringing about equilibrium, with lower values of these constants corresponding to larger equilibration times. For detailed expressions of $f^{eq,u}$, $f^{eq,t}$ and τ_{coll}, the reader is referred to P. J. O'Rourke et al [3].

Numerical method for integrating the collision terms

The calculation of collisions is split from that of the remainder of the terms in the transport equation, Eq. (1), and the rate of change of f from collisional effects is

$$\frac{\partial f}{\partial t} = \frac{f^{eq,u} - f}{\tau^{u}} + \frac{f^{eq,t} - f}{\tau^{t}} \tag{3}$$

A discrete computational particle method is used to solve for the particle distribution rather than direct solution of the Liouville equation. The numerical method for integrating the BGK-like collision terms in Eq. (3) is performed within the framework of the MP-PIC method, and the total numerical solution of the solid-gas phase is solved in framework of the CPFD® numerical method. The detailed numerical equations and solution for the particle collision model are documented in P. J. O'Rourke et al. [3] and an overview of the CPFD® solution method is given in reference [6].

Results and discussion

The liquid injection experiment by Ariyapadi et al. [8] is used to test the collisional exchange model. Ariyapadi et al. used a non-intrusive x-ray imaging technique to investigate the horizontal injection of liquid and gas-liquid jets into a gas-solids fluidized bed. By examining the x-ray images of the fluidized bed, they obtained jet penetration lengths and expansion angles under varied injection conditions. X-ray movies showed the transient jet movement and the interaction between the liquid jet and the fluidized bed. The experiment was carried out in a fluidized bed with a cross section of 20 cm by 20 cm and a height of 100 cm. A straight-tube injection nozzle was horizontally inserted into the bed and located about 4 cm above the air distributor plate at the bottom. Horizontally the nozzle exit (injection point) is 4 cm from the left wall and 10 cm from the front wall. The injection direction is from left to right (Figures 1(c) and (f)). Experimental conditions are listed in Table 1 together with calculation conditions.

Because the experimental results presented by Ariyapadi et al. focused primarily on liquid jet penetration, the numerical simulation in this study focuses on jet penetration lengths under different injection and fluidization conditions. Calculated penetration of liquid into the bed is compared with experimental data. Two sets of injection and fluidization conditions were simulated, these conditions correspond to Tests L1 and L3 in Ariyapadi et al. [8] paper, and parameters are listed in Table 2. The liquid mass flow rate and jet velocity in Test L1 are much smaller than those in Test L3 while the fluidization gas velocity is larger in Test L1.

To match the experimental pre-injection conditions, a calculation with fluidizing gas velocity of 4.2 cm/s was first run to 24 s without liquid injection. At this time the bed was fully fluidized and reached a quasi steady condition. Figures 1(d) and (e) show the particle volume fraction distribution in the whole fluidized bed and in a thin, center-cut slice at 24 s just before liquid injection (simulation time is displayed at the top of the snapshots). Liquid injection with conditions of Test L3 was turned on at 24 s and lasted 10 s with a total calculation time of 34 s. Figure 1(f) shows the liquid distribution at the end of the calculation (34 s). The red color shows the liquid fraction defined as the ratio of the liquid mass to the total particle mass with bright red being 100% liquid. The liquid fraction for pure liquid is 1, and the liquid fraction for a dry particle is 0, independent of their size. For wetted particles, the liquid fraction falls between 0 and 1.

The calculation for Test L1 was performed by restarting the previous calculation at 24 s with the fluidizing gas velocity changed to 5.6 cm/s (from 4.2 cm/s). The calculation was run for 6 seconds for the fluidized bed to reach steady condition at the new fluidization velocity. At 30 s, liquid injection with conditions of Test L1 was turned on and lasted 10 s. Figures 1(a) and (b) show the pre-injection fluidized bed, and Figure 1(c) shows the liquid fraction at the end of the calculation.

Ariyapadi et al. [8] defined the jet penetration as the maximum average horizontal distance from the nozzle tip that the injected liquid penetrates into the bed. In the movies of the experiment, the jet penetration length was observed to fluctuate with time in the bubbling bed. The reported penetration length for a particular test case was obtained by averaging the instantaneous penetration lengths from several movie-images acquired at different injection times. The simulation jet penetration length is calculated the same way by averaging the instantaneous penetration at several times, the latter being the length of the liquid spray path before it is significantly deflected or dispersed. Figure 2 shows time snapshots of the liquid-particles only (solid and partially wetted solid particles are not shown). Under each of the snapshots is the instantaneous jet penetration length. It should be noted that, similar to the experiment, the measurement of the jet penetration length is not accurate because after the initial jet penetration, liquid-particles may be advected by the bed motion (fluidizing gas and particle motion), and older injected liquid-particles may return to the injection path .

Table 1. Experimental and Barracuda simulation parameters for liquid injection into a fluidized bed

	Ariyapadi et al. [8] (Section-I)	Barracuda Simulation
Bed dimensions	20 × 20 × 100 cm	20 × 20 × 65 cm
Solid particle density	(FCC) 1500 kg/m^3	1500 kg/m^3
Solid particle size distribution	Standard FCC (Sauter mean diameter ~ 70 μm)	Standard FCC
Bed inventory	~ 9 kg	9.3 kg
Expanded bed height	~ 40 cm	~ 40 cm
Injected liquid	Ethanol (95 % v/v)	Ethanol
Density of injected liquid	~ 800 kg/m^3	800 kg/m^3
Nozzle diameter	0.82 mm	Point source injection
Liquid droplet diameter	Not reported	0.42 mm
Injection time	10 s	10 s
Fluidization gas/conditions	Air at ambient conditions	Air at ambient conditions
K_u	N/A	0.05
K_t	N/A	0.02

Table 2. Liquid jet parameters for Tests L1 and L3 by Ariyapadi et al. [8]

	Test L1	Test L3
Liquid mass flow rate (g/s)	2.9	6.0
Jet speed (m/s)	6.6	13.9
Fluidizing gas velocity (cm/s)	5.6	4.2

The averaged jet penetration lengths for the two test cases are listed in Table 3. The calculated jet penetration predicted by Barracuda is somewhat longer than that reported from the experiments. Both the experiment and simulation show that the jet penetration in Test L1 is much shorter than that in Test L3. The reduced penetration in Test L1 is from a a smaller liquid flow rate and jet speed and a higher fluidization velocity which results in a stronger cross-flow dispersing the jet.

Table 3. Jet penetration length by experiment and simulation

Case	Experiment by Ariyapadi et al. [8]	Current simulation
Test L1	3.3 cm	5.1 cm
Test L3	7.3 cm	9.7 cm

Concluding remarks

The penetration of liquid into a fluidized bed and the subsequent distribution of liquid from solid to solid is calculated. The addition of the BGK-type approximation for collision transport extends the MP-PIC method for describing gas-liquid-solid bed dynamics. The calculation of jet penetration compared reasonably with experimental data. The accuracy of mixing and transport of liquid between solids could not be assessed from the experiment, but the predicted transport of liquid between solids and dispersion within the bed appears reasonable. The model is included in a generalized form within the larger framework of the Barracuda CPFD® code.

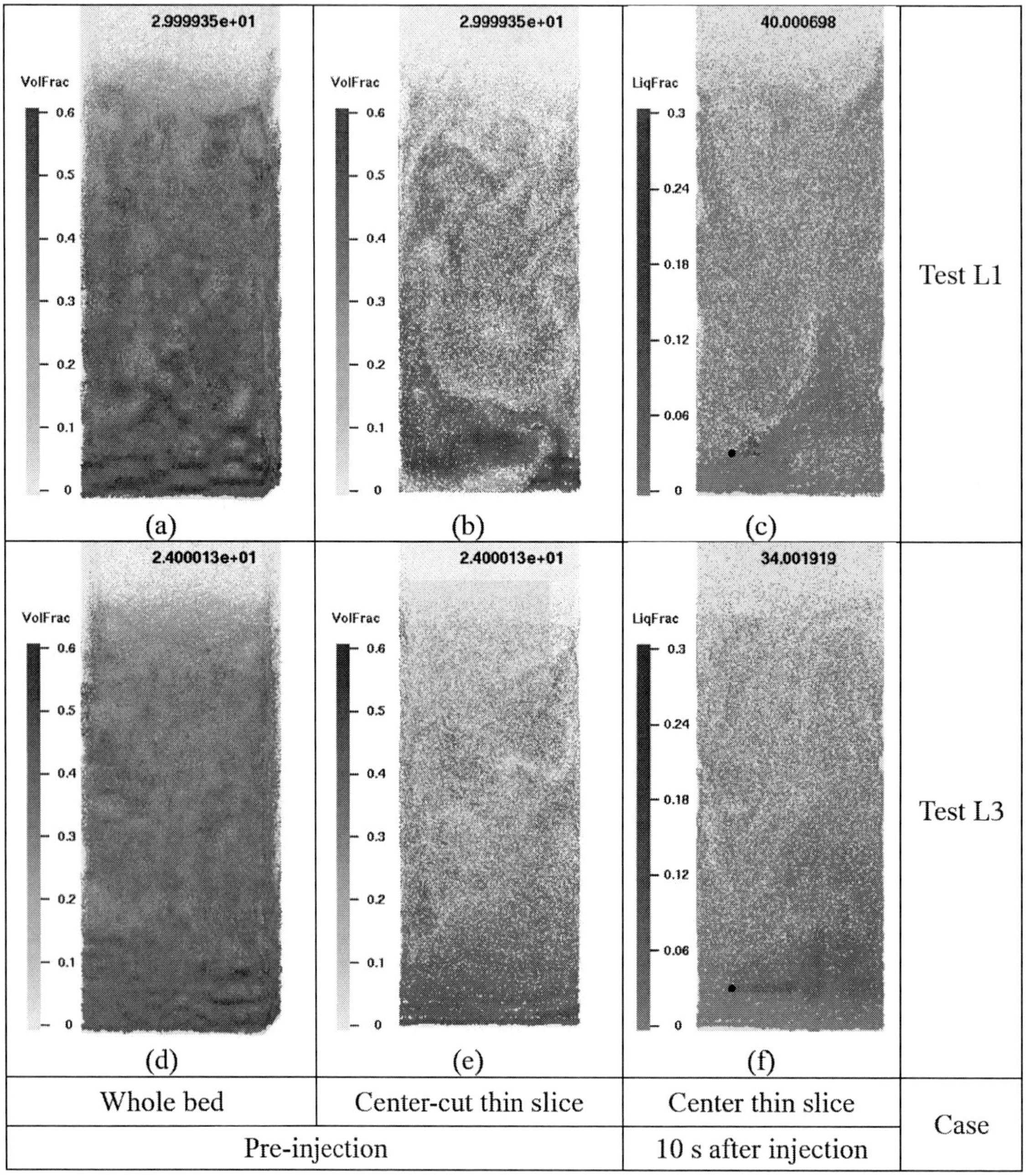

Figure 1. Snapshots of the fluidized bed before and after liquid injection. Liquid injection starts at 30 s for Test L1 and 24 s for Test L3. In both cases injection lasts 10 s. First column: whole fluidized bed prior to liquid injection with particles colored (gray-scale) by volume fraction. Center column: Thin center-slice showing particle colored (gray-scale) by volume fraction prior to liquid injection . Third column: Thin center-slice showing particles colored (gray to red) by the fraction of liquid film at 10 s after the injection. The black dot shows the injection location.

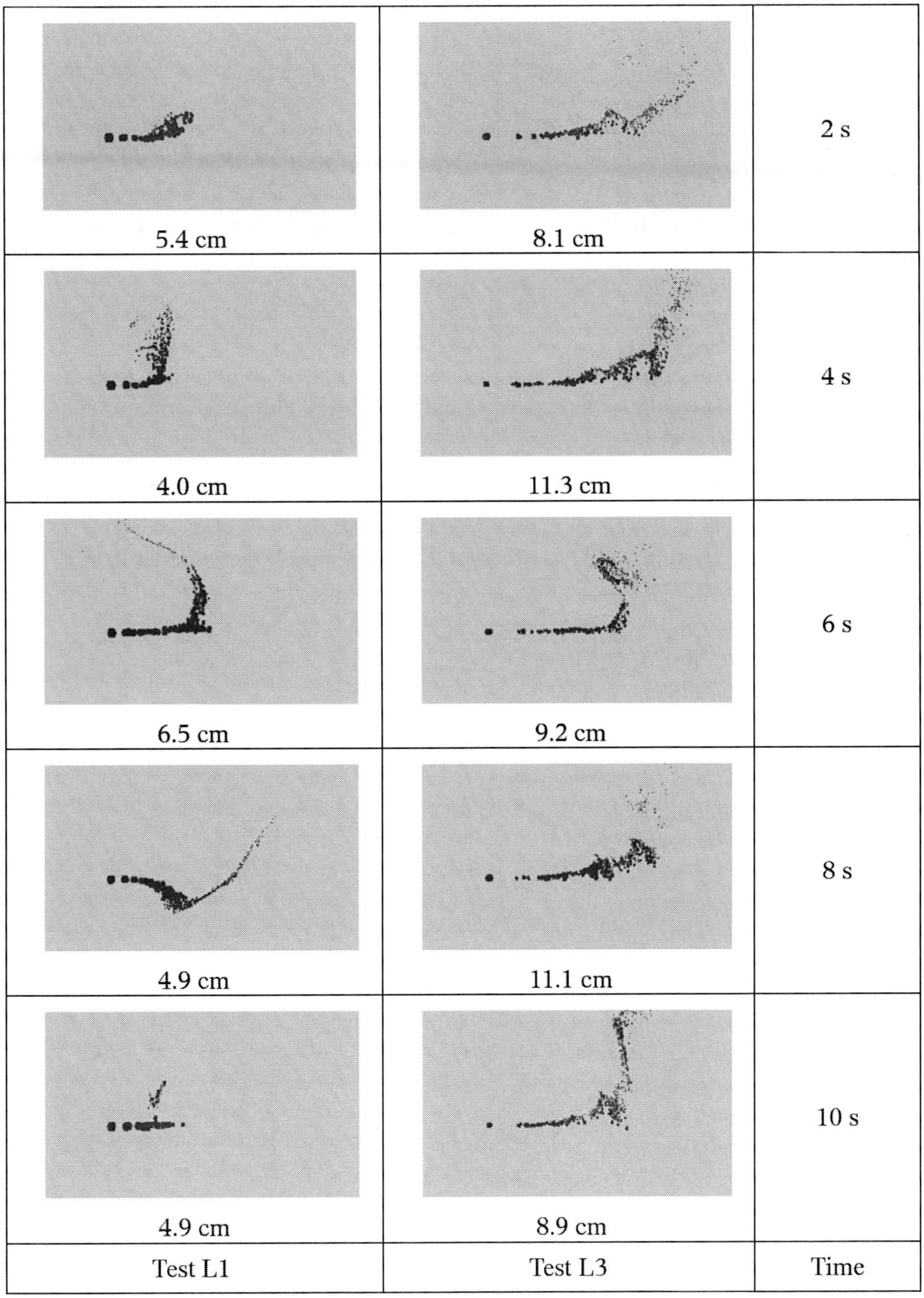

Figure 2. Injected liquid droplet distribution at different injection time showing the jet penetration for two sets of calculation conditions. Partially wetted solids and dry solids are not shown. The column to the right shows time after injection. The numbers in center-meters denotes the penetration lengths.

Reference

1. X. Song, et al., "Hydrodynamics of the Reactor Section of Fluid Cokers," *Powder Technology* **147**, 126-136 (2004).

2. M. Gray, et al., "Kinetics of Cracking and Devolatilization during Coking of Athabasca Residues," *Ind. Eng. Chem. Res.* **43,** 5438-5445 (2004).

3. P. J. O'Rourke, P. Zhao and D. M. Snider, "A Model for Collisional Exchange in Gas-Liquid-Solid Fluidized Beds", In preparation.

4. M. J. Andrews and P. J. O'Rourke, "The Multiphase Particle-in-Cell Method (MP-PIC) Method for Dense Particle Flow," *Int. J. Multiphase Flow* **22**, 379-402 (1996).

5. D. M. Snider, P. J. O'Rourke, and M. J. Andrews, "Sediment Flow in Inclined Vessels Calculated Using Multiphase Particle-in-Cell Model for Dense Particle Flows," Int. J. Multiphase Flow 24, 1359 (1998).

6. D. M. Snider, "An Incompressible Three-Dimensional Multiphase Particle-in-Cell Model for Dense Particle Flows," *J. Comput. Phys.* **170**, 523-549 (2001).

7. W. G. Vincenti and C. H. Kruger, Jr., *Introduction to Physical Gas Dynamics* (Robert E. Krieger Publishing Company, Huntington, NY, 1975).

8. S. Ariyapadi, D. Holdsworth, C. Norley, F. Berruti and C. Briens, "Digital X-ray Imaging Technique to Study the Horizontal Injection of Gas-Liquid Jets into Fluidized beds," Int. J. Chem. Reactor Engineering **1**, Article A56 (2003).

Kinetics of Noncatalytic Water Gas Shift Reaction In a Supercritical Water Medium

Jason Picou, Jonathan Wenzel, Missouri University of Science and Technology, H. Bryan Lanterman, DRS Technical Services, Inc., Alexandria, VA and Sunggyu Lee, Missouri University of Science and Technology, Rolla, MO

Abstract

The kinetics of the noncatalytic water gas shift reaction in a supercritical water medium was investigated at various temperatures in a specially designed 0.4-L Haynes Alloy 230 tubular reactor. Experiments were performed noncatalytically in a continuous mode of operation at a constant pressure of 24.1 MPa, water to CO ratio of 22 to 1, and at temperatures varying from 770 to 1050 K. The reactor effluent gas consisted of H_2, CO_2 and CO, with the CO concentration decreasing with increasing temperature. The CO concentration in the effluent gas reached a minimum of 2% at 1050 K. Using global first order kinetics of the type $k' = A \cdot \exp(-E_a/RT)$ and provided that water was in excess so as to negate the influence of water concentration on the rate law, the kinetic parameters for the rate equation for CO conversion were determined to be: $A = 2.512 \times 10^5\ s^{-1}$ and $E_a = 1.325 \times 10^5$ J/mol•°K.

Introduction

The water gas shift (WGS) reaction is a reversible, exothermic reaction between carbon monoxide and water to produce carbon dioxide and hydrogen.

$$CO + H_2O \leftrightarrow CO_2 + H_2 \qquad \Delta H^\circ_{298} = -41.2\ \text{kJ/mol} \qquad (1)$$

The water gas shift reaction is an important industrial reaction, used in the catalytic converters of automobiles, occurring alongside other reactions during catalytic steam reformation of hydrocarbons, as a parallel reaction to the methanol synthesis reaction from syngas, and as a separate process after gasification and pyrolysis reactions.[1] Whenever the water gas shift reaction is used as a separate process, it is usually catalyzed by either a Fe_2O_3/Cr_2O_3 or a Cu/ZnO catalyst, depending on the temperature. The two catalysts are used in two reactors in series if a substantial or total reduction in carbon monoxide is desired.[2] The reaction is normally conducted at temperatures of 625 to 875 K for the high temperature shift, and 425 to 575 K for the lower temperature shift conversion, with pressures as high as 3 MPa.[3] The reason for using two reactors at two temperatures is because exothermic reversible reactions have faster kinetics at higher temperatures, but are limited by equilibrium conditions. At lower temperatures, the reaction is slower and requires a larger reactor, but higher conversion is achieved.[4] Figure 1, the Gibbs free energy of reaction as a function of temperature, shows that the forward reaction is unfavorable at temperatures above 1100 K.[5] Carbon monoxide is a catalyst poison for proton exchange membrane (PEM) fuel cells, and must be thoroughly removed using the water gas shift reaction as well as other separation means. Various studies have been published regarding the non-catalyzed water gas shift reaction in supercritical water, however most of these studies[6,7,8,9] were conducted at lower temperature ranges of the supercritical water domain. As such, there is a dearth of kinetic information regarding the direct conversion of carbon monoxide in a supercritical water medium, especially at high temperatures far above the critical temperature of water, i.e., 374°C.

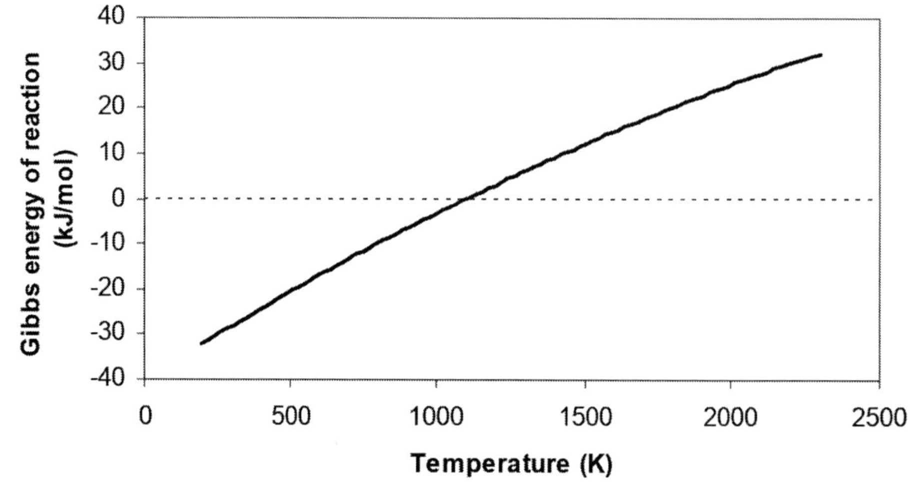

Figure 1. The Gibbs free energy change for the water gas shift reaction as a function of temperature.[5]

Apparatus

The water used for this work was deionized water, and the carbon monoxide was Airgas compressed carbon monoxide with a purity of 99.5%. The supercritical water system consists of a liquid feed system, integrated heat exchanger, feed preheat, carbon monoxide feed system, supercritical water reactor, reactor heaters, sample collection system, and data acquisition and control system, of which a schematic process flow diagram is illustrated in Figure 2. The supercritical water reactor has a volume of 383 mL, and is constructed of Haynes Alloy 230, which is an alloy of mostly nickel, chromium, tungsten, and molybdenum, among with other elements.[10] Analysis of the gaseous effluent was performed using a HP 5890 Series A gas chromatograph with a thermal conductivity detector. The gas chromatograph is calibrated to detect H_2, N_2, CO, CH_4, CO_2, C_2H_2, C_2H_4, and C_2H_6. The carrier gas for the gas chromatograph was Airgas Ultra-pure Carrier Grade Argon with a purity of 99.9995%.

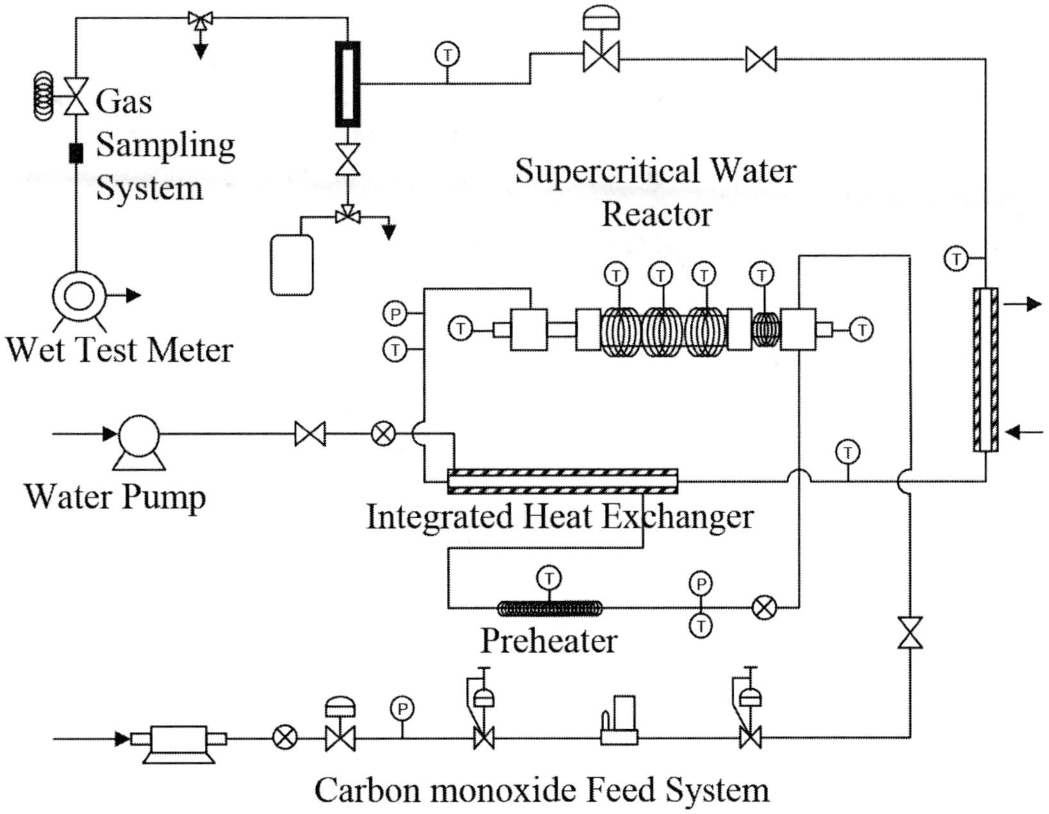

Figure 2. A schematic of the supercritical water reaction system at MS&T

Experimental

Assuming that the rate law for the water gas shift reaction is a simple power-law rate law, of the form

$$-\frac{dC_{CO}}{dt} = kC_{CO}^{\alpha}C_{H_2O}^{\beta} \tag{2}$$

where $-r_{CO}$ is the rate of disappearance of carbon monoxide, C_{CO} and C_{H_2O} are the reactor concentrations of carbon monoxide and water, respectively, and k is the rate constant. It is assumed that the forward water gas shift reaction follows global first-order kinetics, hence $\alpha = 1$. Since all experiments conducted have an excess of water, the water concentration is considered constant and the rate equation reduces to

$$-\frac{dC_{CO}}{dt} = k'C_{CO} \tag{3}$$

Further experiments would have to be conducted to determine the effect of water concentration on the water gas shift reaction in order to regress α and β and the rate constant k. In order to determine the rate coefficient, k', the temperature of the reaction was varied between 770 to 1050 K. A constant pressure of 24.1 ± 0.14 MPa, a water flow rate of 14.7 ± 0.9 grams per minute and a carbon monoxide flow rate of 1.1 ± 0.01 liters per minute was maintained for all experiments. This flow rate corresponds to a molar water to carbon monoxide ratio of 22:1. The space time was calculated, as a function of inlet fluid density, using the Ping-Robinson equation of state with Van der Waals' mixing rules. The space time decreases for increasing temperature because as temperature increases the molar volume increases, which in a constant volume reactor at a constant feed rate decreases the space time. Table 1 outlines the temperature and space time for each experimental condition.

Table 1. Experimental run matrix using carbon monoxide in supercritical water. Pressure, water and carbon monoxide flow rate are kept constant while temperature is varied.

Run ID	Temperature (K)	Space time (sec)
5	768	123
6	818	107
1	867	104
7	893	90.1
2	916	91.7
3	984	91.3
4	1048	78.7

Results and Discussion

Upon performing the experiments in Tables 1, the effluent gas concentration and flow rate were measured, the results of which are illustrated in Table 2. No other gaseous species other than hydrogen, carbon monoxide and carbon dioxide were detectable. From the data, the conversion of carbon monoxide was calculated, and is illustrated as a function of temperature in Figure 3. Carbon monoxide conversion was defined as the outlet flow rate of carbon dioxide divided by the inlet flow rate of carbon monoxide.

Table 2. Summary of gaseous effluent concentration and flow rate.

Run ID	Effluent gas flow rate (slpm)	Gas composition (mol %)		
		H_2	CO	CO_2
5	1.2	1.7	94.5	1.7
6	1.3	4.9	88.7	4.4
1	1.4	15.6	67.5	14.9
7	1.6	27.1	45.0	25.9
2	1.9	40.1	19.5	38.4
3	2.4	47.5	3.8	46.6
4	2.3	48.1	2.0	47.9

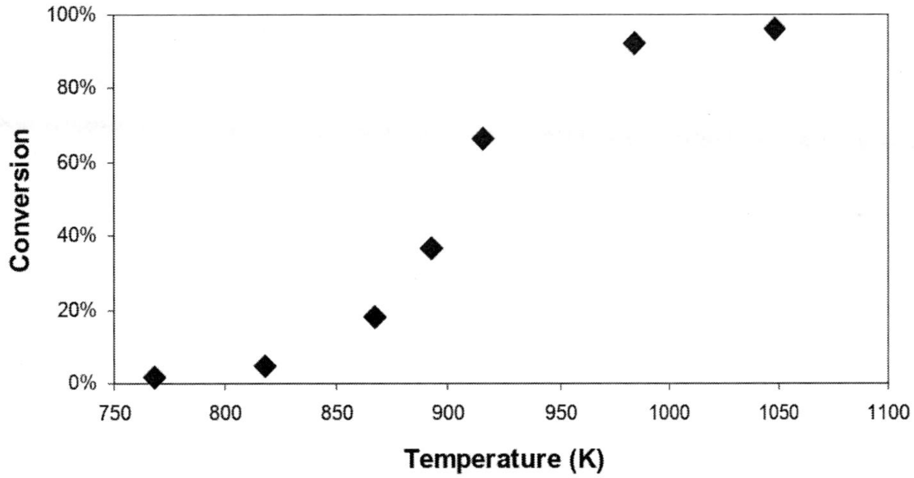

Figure 3. The carbon monoxide conversion as a function of temperature

Upon integration of Equation 3, the resulting equation is

$$C_{CO_1} = C_{CO_0} e^{-k't} \quad (4)$$

where C_{CO_1} is the outlet concentration of carbon monoxide, C_{CO_0} is the inlet concentration, t is the space time in seconds for the experiment and k' is the rate constant in s^{-1}. The Arrhenius equation is defined by the following relation:

$$k' = Ae^{-E_a/RT} \quad (5)$$

where A is the frequency factor, E_a is the activation energy, R is the universal gas constant and T is the temperature in Kelvin. When the natural log of the rate constant is plotted against the inverse of the temperature for each experiment, the slope is equal to $-E_a/R$ and the intercept is equal to the natural log of A. Figure 4 is an Arrhenius plot of the experiments conducted above. From this plot, the frequency factor (A) is determined to be 2.51×10^5 s^{-1}, or $10^{5.4}$ s^{-1}, and the activation energy (E_a) is 132.5 kJ/mol for the non-catalytic supercritical water gas shift reaction. Other studies of the non-catalyzed water gas shift reaction in supercritical water have determined activation energies between 95 to 145 kJ/mol and between $10^{3.3}$ s^{-1} to $10^{7.2}$ s^{-1} for the frequency factor.[6, 7, 8, 9]

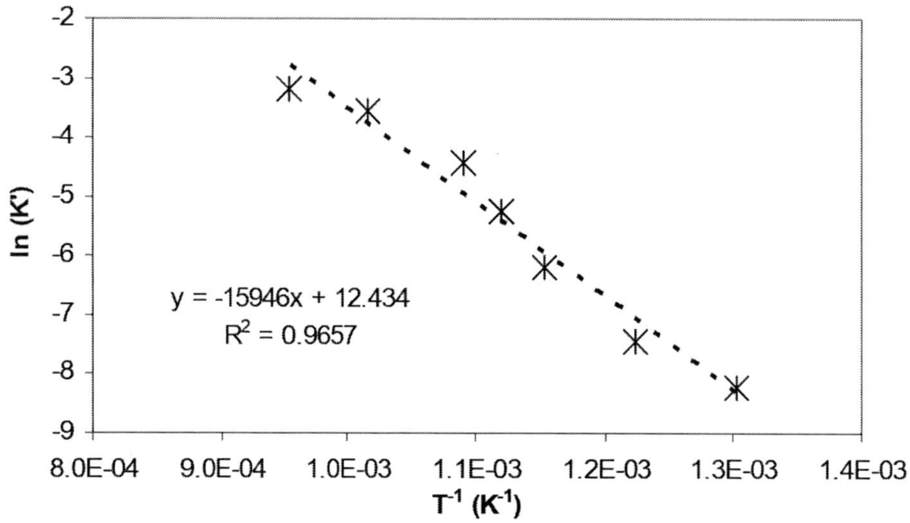

Figure 4. Arrhenius plot for the non-catalytic water gas shift reaction in supercritical water.

Conclusion

The forward water gas shift reaction was examined in a Haynes Alloy 230 supercritical water reactor without the use of a heterogeneous catalyst. Kinetic parameters for the forward water gas shift reaction were calculated for experiments conducted at 24.1 MPa, 770 to 1050 K, and at a 22:1 water to CO molar ratio. Assuming that the reaction follows global first order kinetics, and conducting all experiments with an excess of water so as to negate the influence of water concentration on the rate law, the rate constant *k'*, activation energy and frequency factor were regressed. Further experiments will have to be conducted to evaluate the effects of varying water concentrations upon the forward water gas shift reaction in supercritical water. For the reaction conditions examined in this study, the rate of reaction increases as the temperature increases, with the maximum conversion of carbon monoxide to carbon dioxide, 96%, occurring at the highest temperature, 1050 K. The Arrhenius activation energy was found to be 132.5 kJ/mol with a frequency factor of $10^{5.4}$ s^{-1}.

Acknowledgements

The current project was sponsored by the U.S. Army under PE Number 0602705A through a subcontract from DRS Technical Services, Inc. The authors are grateful for insightful technological guidance and supervision provided by Dr. Terry Dubois, CERDEC, U.S. Army, Ft. Belvoir, Virginia.

References

(1) Lee, S.; *Alternative Fuels*, Taylor & Francis, Philadelphia, 1996; pp140-157.
(2) Flytzani-Stephanopoulos, Maria; Qi, Xiaomei; Kronewitter, Scott; *Water-gas Shift with Integrated Hydrogen Separation Process*; Final Report to DOE on Award DE-FG2600-NT40819; Tufts University: Medford, MA, February 2004 pp. 3
(3) Callaghan, Caitlin A. Kinetics and Catalysis of the Water-Gas-Shift Reaction: A Microkinetic and Graph Theoretic Approach. Ph.D. Dissertation, Worcester Polytechnic Institute, Worcester, MA,

March 2006 pp 40-43, 117
(4) TeGrotenhuis, W.E.; King, D.L.; Brooks, K.P.; Golladay, B.J.; Wegeng, R.S.; Optimizing Microchannel Reactors by Trading-Off Equilibrium and Reaction Kinetics through Temperature Management. Presented at the AIChE Spring National Meeting, New Orleans, LA, March 2002.
(5) M. W. Chase, Jr., C. A. Davies, J. R. Downey, Jr., D. J. Frurip, R. A. McDonald, and A. N. Syverud, *JANAF Thermochemical Tables*, 4th ed. [J. Phys. Chem. Ref. 1998]
(6) H.R. Holgate; P.A. Webley; J.W. Tester; Carbon monoxide oxidation in supercritical water: the effects of heat transfer and the water-gas shift reaction on observed kinetics. *Energy Fuels* **1992**, *6*, 586-597.
(7) S.F. Rice; R.R. Steeper; J.D. Aiken; Water density effects on homogeneous water-gas shift reaction kinetics. *J. Phys. Chem. A* **1998,** *102*, 2673-2678.
(8) T. Sato; S. Kurosawa; T. Adschiri; K. Arai; Kinetics of water gas shift reaction in supercritical water. *J. Chem. Eng. of Japan* **1999**, *37* (3), 443-448.
(9) T. Sato; S. Kurosawa; R.L. Smith Jr.; T. Adschiri; K. Arai; Water gas shift kinetics under noncatalytic conditions in supercritical water. *J. Supercritical Fluids* **2004**, *29*, 113-119.
(10) *Haynes® 230 Alloy*, Haynes High-Temperature Alloys International, 2004 pp. 2
(11) Fogler, H.S.; *Elements of Chemical Reaction Engineering, 3rd Ed.*, Prentice Hall, New Jersey, 1999
(12) Smith, J.M.; *Chemical Engineering Kinetics, 3rd Ed.*, McGraw-Hill, New York, 1981.

HIGH THROUGHPUT OF NON-STEADY-STATE CATALYTIC ACTIVITY CHARACTERISTICS USING TEMPORAL ANALYSIS OF PRODUCTS (TAP) AIChE 2008 SPRING NATIONAL MEETING

Gregory S. Yablonsky, Saint Louis University, St. Louis, MO, USA
Rebecca Fushimi, Washington University in St. Louis, St. Louis, MO, USA
John T. Gleaves, Washington University in St. Louis, St. Louis, MO, USA
Xiaolin Zheng, Washington University in St. Louis, St. Louis, MO, USA
Renato Feres, Washington University in St. Louis, St. Louis, MO, USA
Denis Constales, Ghent University, Ghent, Belgium
Alex Mueller, Washington University in St. Louis, St. Louis, MO, USA

Introduction

The enormous wealth of kinetic and structural information from model and practical catalytic system studies has greatly increased our understanding of the link between catalyst structure and performance. For a number of important practical systems, careful kinetic studies using steady- and non-steady-state techniques have been performed yielding rate parameters, and important details of the catalytic reaction mechanism. For a number of model systems, combined surface science and kinetic studies have provided the direct link between surface structure and performance.

The development of practical catalysts such as supported metals or metal oxides often involves a multi-step bulk preparation procedure followed by surface activation. Many practical catalysts contain more than one transition metal, which may exist in different oxidation states. Notwithstanding significant advances in understanding fundamental catalytic processes, practical catalyst development is hampered by a lack of fundamental information relating the surface composition of the catalytic phase and catalyst performance. The surface of a practical catalyst is likely to be compositionally different from the bulk, structurally and compositionally non-uniform, and may change during the course of a catalytic reaction. As a result, information gleaned from determining the bulk crystal structure of a catalyst is of limited value in characterizing a catalyst surface. The chemical properties of a surface may also be strongly influenced by various adspecies, which in turn may change during the course of a reaction. Adspecies, reaction intermediates, and adsorbed reaction products often have short surface lifetimes, and may exist in low concentration on the surface. As a result, they are difficult to detect and characterize under reaction conditions. Also, the catalytic sites on a practical catalyst may occupy only a fraction of the total catalyst surface making their identification difficult even with surface sensitive techniques. These factors pose serious obstacles to the direct observation and characterization of catalytic sites, and as a result it is difficult to develop detailed structure-activity correlations on practical catalysts.

This paper presents a new approach to high throughput monitoring of non-steady-state catalytic activity characteristics based on the combination of catalyst preparation, which combines in a single apparatus an atomic beam deposition (ABD) system with a temporal analysis of products (TAP-2) reactor system [1], and a new methodology of non-steady-state kinetic characterization. This approach is focused on establishing direct, reproducible correlations between changes in surface composition and changes in catalyst activity.

Catalyst samples described in this paper were prepared by directly adding metal atoms and metal oxide clusters in submonolayer amounts to the surface of micron-sized particles (catalysts or support particles) using atomic beam deposition (ABD). The ABD technique is demonstrated by the deposition of different amounts of Pd and PdO on quartz particles, and the reactivity of the resulting materials toward the CO oxidation reaction was characterized using TAP pulse response and pulse temperature-programmed reaction experiments.

Experimental

Atomic Beam Deposition (ABD) on Micron-Sized Particles

A key challenge for direct atomic deposition on particles is obtaining uniform coverage. Atomic beam deposition is a line-of-sight process so that the only surface to receive coverage is one directly in line with the atomic beam. As a result, a static particle arrangement, with surfaces not in line with the beam will not receive uniform coverage. Our approach involves tumbling a bed of particles in an atomic beam so that, over a period of time, every particle surface is randomly exposed to the beam. Our atomic beam deposition system consists of a vacuum chamber containing a sample holder and metal target, and a laser for ablating the target surface to produce metal atoms (Fig. 1).

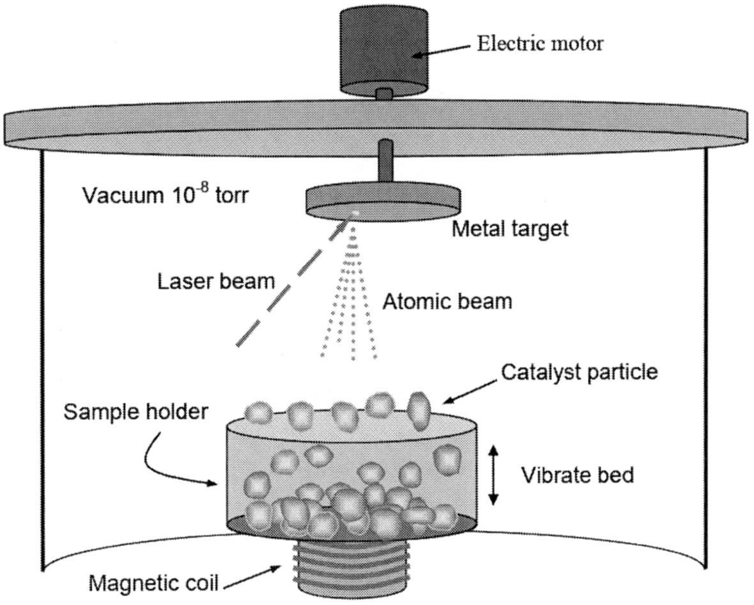

Fig. 1. Simplified schematic of atomic beam deposition experiment.

A pulsed laser beam ablates the surface of a rotating metal target sending a plume of atoms toward a vibrating particle bed. The sample holder contains a magnetic diaphragm in its bottom. When a pulsing current is sent through the magnetic coil the diaphragm vibrates and the particles bounce around so that all particle surfaces are equally exposed to the beam.

The sample holder is a shallow stainless steel dish, one inch in diameter, with a magnetic diaphragm in the base. A typical loading would consist of a single layer of particles 100 to 350 microns in diameter. This particle size range is ideal for TAP Knudsen pulse response experiments.

An electromagnet is positioned beneath the sample holder, and when a pulsing current is sent through the electromagnetic coils, it causes the magnetic diaphragm to vibrate. The vibrating diaphragm causes the particles to vibrate and tumble as well. Changing the frequency and amplitude of the current passing through the electromagnet changes the speed and intensity at which the particles tumble. The sample holder is positioned beneath a flat metal target, which is continuously rotated during an experiment. Metal atoms are produced by focusing a beam of laser light from a pulsed excimer laser (Lambda-Physik EMG-100, XeCl active medium, 150 mJ pulse energy) onto the surface of the metal target. The laser pulse ablates the metal surface and creates a plume of atoms. The number of metal atoms produced per laser pulse is a function of the pulse intensity, fluence, and pulse duration. An important advantage to using an excimer laser is that it produces very narrow pulses ($\approx$20 nanoseconds) in the UV region of the spectrum, and is very effective at producing metal atoms.

Typical pressures in the vacuum chamber during deposition are $< 10^{-6}$ torr. At this pressure the mean free path is $\approx$ 50 m so that in the space between the target and sample holder gas phase collisions can be neglected. Temperature control of the microparticles is independent of the temperature of the deposition source. The bed temperature can be maintained at a value where activated chemical interactions between the deposited species and the substrate particles are negligible. In this case, surface reactions can be experimentally decoupled from the deposition process, and the compositional and structural evolution of a surface and its chemical properties can be followed in time.

Catalyst Kinetic Characterization

The catalytic properties of deposition samples were measured in a TAP-2 reactor system. A detailed description of the TAP-2 apparatus and experimental methodology can be found in papers [2-3]. The TAP-2 system has the ability to perform vacuum pulse response experiments and atmospheric pressure experiments, including steady-state, step-transient, and temperature programmed experiments. The TAP-2 reactor consists of three main parts: a pulse valve/manifold assembly, a microreactor, and vacuum chamber with a mass spectrometer (Fig. 2). The pulse valve/manifold assembly contains four high speed pulse valves and a continuous flow valve. The pulse valves can be adjusted to inject very small ($\approx 10^{-10}$ to 10^{-8} mol/pulse), narrow (< 500 μs) gas pulses into the microreactor. The microreactor outlet is coupled directly to the vacuum chamber and can be operated at vacuum conditions or at atmospheric pressure by moving a slide valve over the reactor outlet. In a vacuum experiment, the reactor effluent flows directly into the ionizer of the mass spectrometer, which samples the gas flow as a function of time. The vacuum chamber is specially designed to give a very high pumping speed so that responses measured at the mass spectrometer accurately reflect gas interactions in the microreactor.

A TAP Knudsen pulse response experiment is performed by injecting a narrow gas pulse of "reactant molecules" into an evacuated microreactor containing a solid sample usually in particle form. The observed characteristic in a TAP experiment is the time dependent gas flow F(t) [mol/s] or [molec/s] that escapes from the exit of the microreactor. The flow dependencies also have integral characteristics that are the moments of the "flow-time" dependencies.

In TAP data analysis, moments have proven to be the effective and straightforward mathematical language for characterizing the flow time-dependence that is characteristic of TAP pulse response curves [2, 3, 5]. Experimentally, the moments are integral characteristics of the observed exit flow time dependencies. Mathematically, the moments of the exit flow, $F_{exit}(t)$, are directly related to its solution in the Laplace domain, $\overline{F}_{exit}(s)$,

$$M_n = \int_0^{+\infty} t^n F_{exit}(t)dt = (-1)^n \left. \frac{\partial^n \overline{F}_{exit}(s)}{\partial s^n} \right|_{s=0} \quad (1)$$

Equation (1) allows the moments to be calculated analytically as a function of the kinetic parameters. The parameters can then be determined as functions of the experimentally measured values of the moments. The zeroth moment is proportional to the number of molecules. Normalized to the number of injected molecules, the zeroth moment of the exit flow is directly related to conversion.

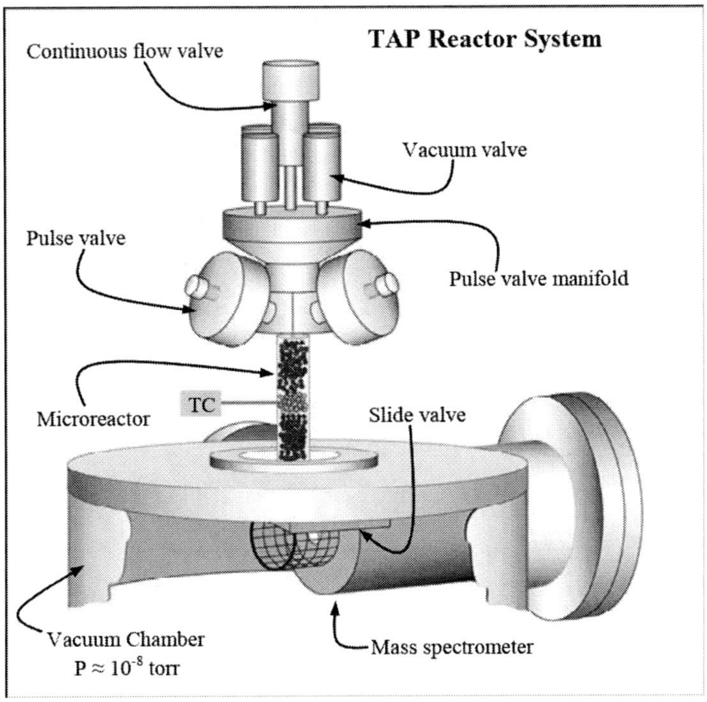

Fig. 2. Schematic of TAP-2 reactor system.

The manifold contains four high-speed pulse valves for TAP vacuum pulse response experiments and a continuous flow valve for atmospheric flow experiments. The microreactor has a diameter of 0.25 inches and a length of 1.65 inches. It contains an internal thermocouple and can be operated isothermally or temperature programmed. A 2-position slide valve assembly separates the reactor outlet from the mass spectrometer and vacuum chamber. In the closed position, the reactor can be operated at atmospheric pressures. The open position is for TAP vacuum experiments.

Results and Discussion

Non-Steady-State Characterization via Pulsed Temperature-Programmed Reaction (TPR) of Palladium (Pd) Catalyst Samples

Immediately following catalyst preparation of $Pd/PdO/SiO_2$ samples, the fresh catalysts were transferred into the TAP-2 reactor system for characterization. To eliminate oxygen acquired during catalyst preparation the $Pd/PdO/SiO_2$ samples were reduced by pulsing CO and ramping the temperature (≈1 °C/s) from 20 to 400 °C. During the temperature ramp the CO, Ar, and CO_2 responses were recorded. Oxygen uptake was performed at low temperature to prevent migration in the catalyst bulk. CO was pulsed while temperature ramp to ensure the removal of added oxygen via reaction rather than migration into the bulk.

Fig. 3 shows CO_2 production as a function of temperature over a deposition sample made with 750 Pd pulses. A maximum in CO_2 production is observed at 154 °C. After the maximum, production drops rapidly as the sample continues to heat. The initial steep drop in CO_2 production is followed by a gradual decrease with periodic bursts giving the appearance of a damped oscillation. All freshly deposited samples exhibited this oscillatory behavior. The reactor temperature was maintained at 400 °C until CO_2 production was < 5% of its maximum value. The trend in CO_2 production and the total amount produced was highly reproducible on separate samples prepared with identical Pd loadings.

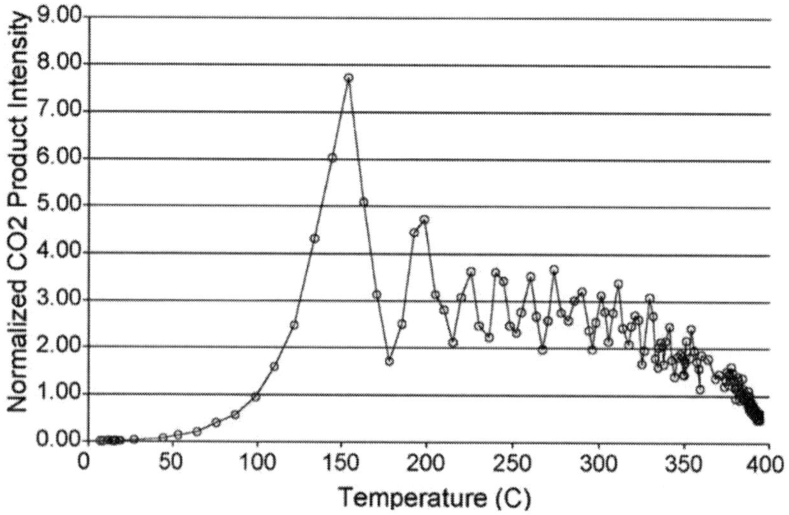

Fig. 3. CO_2 production (determined by measuring the zeroth moment of the CO_2 responses) over fresh catalyst deposit (750 Pd pulses) obtained by pulsing CO and ramping the reactor temperature from 20 to 400 °C.

Heuristically, every peak in the dependence presented in Fig. 3 can be viewed as a typical TPR peak related to a specific form of catalyst oxygen. CO_2 production during the first peak corresponds to the depletion of accessible surface oxygen. Then, the series of peaks that follow can be considered as resulting from reaction of CO with other forms of catalyst oxygen. This, abrupt changes in the amount of oxygen indicate abrupt changes in the Pd/PdO composition. Moreover, the reduction of PdO results in a volume loss of nearly 67% (based on theoretical change in density going from PdO to Pd) and the collapse of the resulting amorphous layer anticipates an atomic reorganization in order to minimize surface energy. Based on the obtained experimental characteristics, the hypotheses on the detailed mechanism and its dependence on the catalyst state can be proposed.

Discussions on the Methodology of Non-Steady-State Catalyst Characterization

State-by-state non-steady-state catalyst TAP characterization has been introduced in the paper by Shekhtman, et al. [6]. In accordance with this procedure, catalyst kinetic characteristics (apparent kinetic constants, substance storages, and time delays) are obtained in the course of the multi-pulse experiments and presented as functions of catalyst composition. Presently, a methodology of our procedure is based on the application of the idea by Shekhtman and Yablonsky (1999) to perform TAP experiments using thin-zone reactors (TZTR). The TZTR is a very useful special case of the three-zone TAP reactor configuration, in which the thickness of the catalyst zone is very small compared to the entire reactor length. Having a very thin catalyst zone makes the change in the gas concentration across the catalyst zone small compared to its average value [7-9]. A key advantage of the TZTR is that the catalyst bed can be changed uniformly by exposing the catalyst to a long series of small pulses at a conversion of up to 80% (it is much higher than in a differential PFR) [10]. Recently (2008), the new configuration of the TZTR has been proposed by Xiaolin Zheng and J. Gleaves et al. [11]. In this case, the reaction zone is collapsed to the surface of a single micron-sized catalyst particle in a bed of inert quartz particles. The particle occupies less than 0.3% of the cross-sectional area of the microreactor, so that the reaction zone can be considered a point source. In a typical experiment, the microreactor was packed with approximately 100,000 inert quartz particles (210-250 μm in diameter) and a single Pt catalyst particle (300-400 μm in diameter) usually positioned in the middle of the reactor bed. An advantage of this configuration is that concentration and temperature gradients in the catalyst bed can be assumed to be negligible for very fast catalytic reactions even in the domain of high conversions. The probabilistic theory of the TAP-experiment over single particle has been developed recently by R. Feres et al. This theory and its application to the TAP-data will be published in our next paper.

Y-Procedure

The method for extracting chemical transformation rate from reaction-diffusion data with no assumption on the kinetic model ("kinetic model-free procedure"), so called the Y-procedure, opens new possibilities for high throughput of the non-steady-state kinetic activity. The mathematical foundation of this procedure proposed by Yablonsky and Constales in 2003 [12] (see in detail [13]) is a Laplace domain analysis of two inert zones in a TZTR followed by transposition to the Fourier domain. When combined with time discretization and filtering, the Y-procedure leads to an efficient method for determining the reaction and reaction rate in the active zone. Using the Y-procedure, the concentration and reaction rate of a non-steady catalytic process can be determined without any pre-assumption regarding the type of kinetic dependence. The Y-procedure is the basis for advanced software for non-steady-state kinetic data interpretation. The Y-procedure can be used to relate changes in the catalytic reaction rate and kinetic parameters to changes in the catalyst surface composition. Particularly, it may provide information about the set of the 'fast' transformations observed within a single pulse. Regarding a set of slow transformations, such information will be provided by the approach based on the moment analysis which presents an evolution of averaged catalyst kinetic characteristics. In the paper, the Y-procedure is illustrated by some examples.

References

1. Fushimi, R., Gleaves, J.T., Yablonsky, G.S., Gaffney, A., Clark, M., Han, S. (2007) *Catalysis Today,* 121(3-4), pp. 170-186.
2. Gleaves, J.T., Yablonskii, G.S., Phanawadee, P., Schuurman, Y. (1997) *Applied Catalysis A,* 160(1), pp. 55-88.

3. Yablonsky, G.S., Olea, M., Marin, G. (2003) *Journal of Catalysis,* 216(1-2), pp. 120-134
4. "The TAP Reactor in Catalysis: Recent Advances in Theory and Practice," (2007) *Catalysis Today,* ed. J. Pérez-Ramírez, E.V. Kondratenko, 121, pp. 1-124.
5. Yablonskii, G.S., Shekhtman, S.O., Chen, S., Gleaves, J.T. (1998) *Ind. Eng. Chem. Res.,* 37(6), 2193-2202.
6. Shekhtman S.O., Yablonsky G.S., Gleaves J.T., Fushimi R. (2003) *Chem. Eng. Sci.,* 58, pp. 4843-4859
7. Shekhtman S.O., Yablonsky G.S., Chen S., Gleaves J.T.(1999) *Chem. Eng. Sci.,* 54, pp. 4371-4377
8. P.Phanawadee, S.O.Shekhtman, Jarungmanorom C., Yablonsky G.S., Gleaves J.T. (2003) *Chem. Eng.Sci.*, 58, pp. 2215-2227
9. Shekhtman S. O., Yablonsky G.S., Gleaves J.T., Fushimi R., *Chem. Eng. Sci.* (2004) 59, pp. 5493-5500
10. Shekhtman, S.O., Yablonsky, G.S. (2005), *Ind. Eng. Chem. Res.,* 44(16), pp. 6518-6522
11. Xiaolin Zheng, Gleaves J.T., Yablonsky G.S., Brownscombe T., Gaffney A, M. Clark, S. Han (2008), "Needle in a Haystack catalysis," *Applied Catalysis A* (accepted for publication)
12. Yablonsky G.S., Constales D., Gleaves J.T. (2002), Systems Analysis Modeling and Simulation, 42, 8, pp. 1143-1166
13. Yablonsky G.S., Constales D., Shekhtman S.O., Gleaves, J.T. (2007) *Chem. Eng. Sci.,* 62, pp. 6754-6767

139g Radiolysis Induced Mobility of Chloride Ions in Zeolite

William David Jacobs and Herbert Lee Nigg. Hydrogen Technology Research Laboratory, Savannah River National Laboratory, Washington Savannah River Company, Mail Drop Building 999-2W, Aiken, SC 29808

The Savannah River Site processes hydrogen isotopes using molecular sieves (zeolites) to dry process streams. The saturated molecular sieves are periodically regenerated to recover the third isotope of hydrogen, tritium. Typically drying or regeneration of molecular sieves does not result in corrosion, but due to the unique beta decay energy of tritium radiolysis induced corrosion has been observed. Zeolite typically has some chloride species as a result of the sol-gel fabrication process. The chloride species are typically assumed fixed and no significant leaching of chlorides is noted. Experiments have been conducted to test the hypothesis that the radiation has contributed to chloride mobility and hence the corrosion of downstream piping.

6th Biennial Electrochemistry Seminar of Iran

The Kinetics of Chlorine Evolution on Dimensionally Stable Anode (DSA) Used in the chlor-alkali Industry

*M.Vahdati. Mohaddes

Petrochemical research & technology company,Iran,Tehran.e-mail:m.vahdati@npc-rt.ir

Abstract

The anodic oxidation of chloride in an acid aqueous media on the mixed metal oxide as a DSA, leads to the chlorine evolution reaction (ClER) [1]. The kinetics and mechanism of the electrooxidation was investigate by linear sweep voltammetry, Tafel slopes and reaction order determinations.

The activation parameters were determined from the ClER rates measured at temperature range of 30-90°C.

The mechanism of the ClER on the DSA is complicated by the specific features of the surface [3], but on the basis of the experimental results, we tried to propose the mechanism of the CLER.

Introduction

The ClER is a well known and apparently plain electrode reaction:

$$2Cl^{-} \leftrightarrow Cl2 + 2e$$

with a positive standard electrode potential (1.36 v). However, three different reaction mechanism are distinguished according to the kinetics and the rate of the separate reaction steps [3]. The basic equations were developed for the volmer – Tafel and the volmer – Heyrovsky routes. The mechanism of the chlorine evolution has been extensively examined e.g. by determination of the tafel slpes.

Kinetic studies have mostly been based on tafel slopes and on reaction orders. The kinetic and mechanism of chlorine electrode reaction varies with the chemical composition and structure of working electrode in aqueous solutions.

Experimental

The typical structure of working electrode (DSA) was used in kinetic studies consists of a flat support (Ti) Coated with a thick layer of metal oxides belongs to permascand company.

The rest potential of the DSA in solution of (H2O + NaCl+HCl) as electrolyte was measured against a saturated calomel electrode (SCE) and was varied out by sweeping potentional between (0.5-1.2 volt) at 0.1 v/s sweep rate. The concentration of chloride ions was ranged between (0.5-5 M) and changes in pH values were performed by addition of HCl and maintained between 2-4, because of in the limited range, the ClER is the main electrode reaction. The activation thermodynamic parameters were determined from the ClER rates measured at temperature arange of (30-90°C).

Results and Discussion

Voltammetric curves, E vs. I were recorded to monitor the Cl2 evolution at lower overpotentional than standard potential. Curves were recorded between (0.5-1.2 v vs. SSCE). Figure 1 shows representative voltammetric profiles recorded before the kinetic experiments.

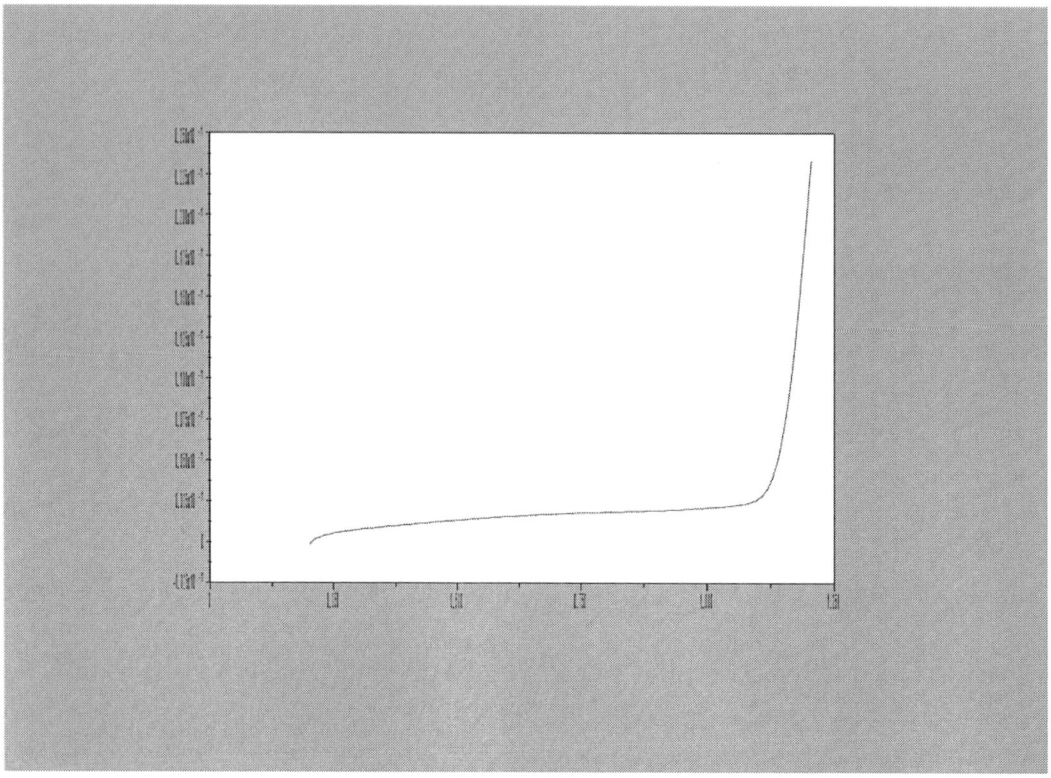

Fig. 1 Typical LSV curve on the DSA working electrode in 2 M NaCl-HCl solution at 0.1 v/s sweep rate.

Reaction rate constant were determined by a conventional method that is following the change of current density as a reaction rate, in constant overpotential (at 1.15 v vs.SCE) against the concentration of chloride, using linear sweep voltammetry technique. An experimental kinetic parameters was obtained with measuring of current density near the equilibrium potential, eliminates the interference due to the porous structure of the electrode.

The anodic voltammetric current density , I (A/cm2) obtained from I vs. E plots in various temperature (25-28°C) and chloride concentration (0.5-5 M). Table 1.

Table 1. Dependence of the current density on [Cl^-] an Temp

85	70	55	40	25	Temp/C°
Current density (A/cm2)					Conc/M
0.00487	0.00345	0.00271	0.00218	0.00144	0.5
0.00942	0.00875	0.00724	0.00567	0.00415	1
0.05120	0.3210	0.01560	0.00998	0.00891	2
0.32100	0.9810	0.07690	0.05910	0.04980	4
0.65200	0.56800	0.52400	0.43200	0.34100	5.3

To propose an electrode mechanism for the ClER two reaction orders must be determined: (i) with respect to Cl^- at constant pH and (ii) with respect to H+ at constant Cl^- concentration. (ionic strength. Etc.)

Chart 1. shows representative Log I vs. Log [Cl^-] profiles in various temperature. Lines show linearity with correlation coefficient between (0.9078-0.9800) that is obtained for DSA as the current density read at E-1.15 V vs. SCE, is plotted as a function of NaCl concentration.

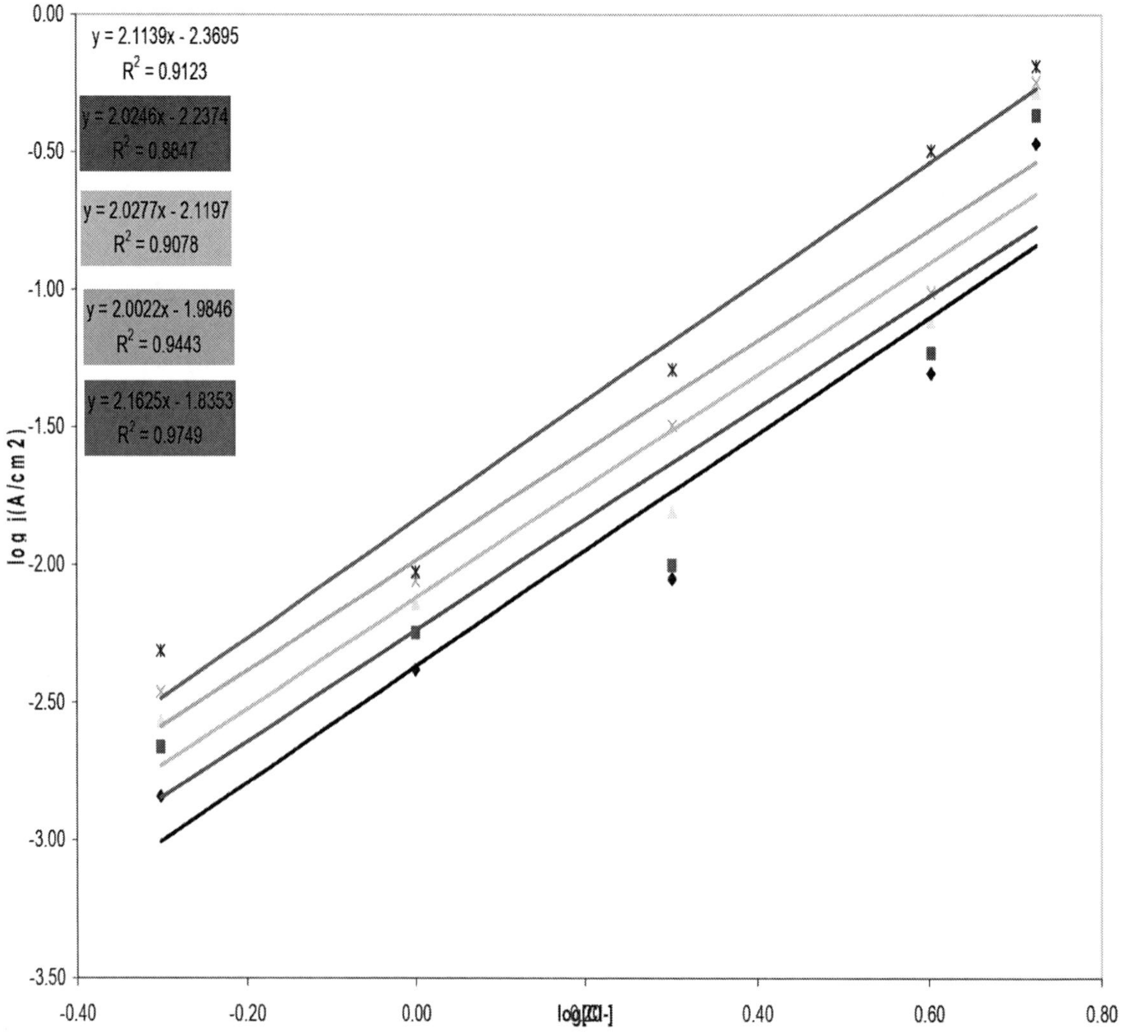

Chart 1. representative Log I vs. Log [Cl^-] profies.

The results reveals a positive influence of the [Cl^-] on the rate of the ClER. The slope of the lines is (2.1 ± 0.1 = υCl^-) with respect to [Cl^-] and rate constant is 8.46E-3 with dimensionally (1/I.t). Rate constant is obtained from rule of rate according to:

$$\text{Log } R = \log k + \upsilon \log [Cl^-]$$

The reaction order with respect to H+ was also determined at E = 1.15 V vs. SCE by changing the HCl concentration between (pH = 1-3). The reaction rate was depressed by increasing acidity but the reaction order with respect to H+ was found to be independent of pH variation in the limited range.

Constant rates are plotted against inverse of temperature in linearity range of temp.

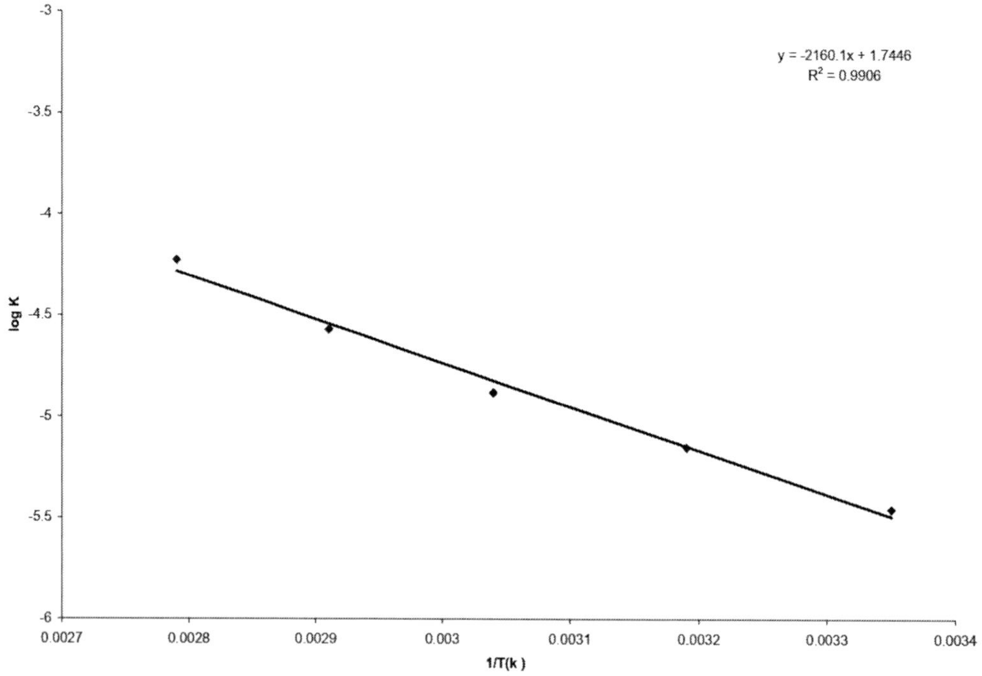

A

ctivation energy Ea = 17.965 kj/mol , A = 5. 73

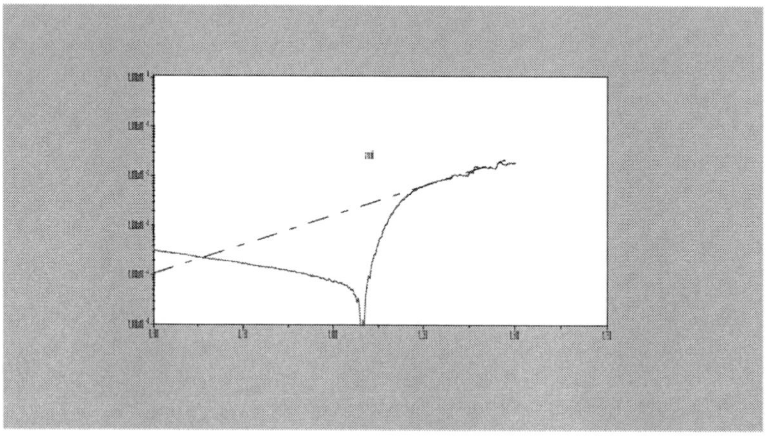

Fig 2. Typical tafel curve on the DSA working electrode in 2 M NaCl-HCl solution at 0.1 v/s sweep rate

Results: 1. Ecorr, obs 1.07 V

2. Ecorr, calc 2.04 V (at E-begin = 1.195 V)

3. icorr 1 A

4. I corr 0.595 A/cm2 (A = 1.68 cm2)

5. Rp (at Ecorr) 3.204 E + 1 ohm

6. Corrosion rate 1.948 E + 1 mm/year

7. ba (tafel slope) 0.352 V, n=1

Reaction mechanism

Several different electrode mechanisms have been proposed for the ClER [4]. Considering the experimental b-values and reaction orders with respect to H+ and Cl^- ions, the mechanism proposed by Erenburg et al. adequately describes the electrode reaction:

b/m V(T = 25°C, $\alpha = 0.5$)

$\equiv S + Cl^- \rightarrow \equiv S - Cl + e$ 120 (a)

$\equiv S + Cl \rightarrow \equiv S - Cl+ + e$ 40 (b)

$\equiv S - Cl{+}{+}Cl^- \rightarrow \equiv S + Cl^2$ 30 (c)

Where $\equiv$ S represents an active surface site.

In agreement with the experimental υH+ value of zero, the above mechanism does not predict any dependency of the reaction rate on pH.

Considering steps (b) or (c) as the rate determining step, rds, reaction orders with respect to Cl^- of 1 or 2 are predicted, respectively.

Considering step (c) as the rds, in wich case an order of reaction with respect to chloride of two and b = 32v is theoretically predicted. In this condition, the reaction rate, I, at $\eta \geq 0.1$ V can be written as : $I = k [Cl-] \exp \{[1+\alpha)F(E-E^\circ)]/R.T\}$

For the exploted potential range, the ClER kinetics on DSA are governed by a single rate determining step. Contrary to the Cl- anions, H+ doesn't exert influence n the electrode kinetics.

Experimental data support the electrode mechanism proposed by Erenbur et al. for the ClER.

References:

1. J. Electrochimica Acta 45 (2000) 2377-2385.
2. J. Braz. Chem. Soc., Vol. 14, No. 3, 388-395, 2003.
3. J. Electrochimica Acta 44 (1999) 2463-2472.
4. J. Electroanal. Chem 228 (1997) 393.

The Kinetics of Chlorine Evolution on Dimensionally Stable Anode (DSA) Used in the chlor-alkali Industry

M.V. Mohaddes and F. Mohammadi

Petrochemical research & technology company, Iran Polymer & Petrochemical Research Institute, 15th km of karaj Highway, Tehran, Iran.

e-mail:m.vahdati@npc-rt.ir

Abstract

The anodic oxidation of chloride in an acid aqueous media on the DSA leads to the chlorine evolution reaction (ClER) [1]. Inour research, we investigate the ClER kinetics by means of voltammetric curves, Tafel lines and reaction order determinations.

The typical structure of electrodes (DSA) used in kinetic studies consists of a flat support (Ti) Coated with a thick layer of metal oxides from permascand company as an industrial electrode.

A convenient method with three electrodes system, is used for measuring the ClER rate [2] that is following the change of current densit in constant overpotential against the concentration of chloride, using linear sweep voltammetry technique with potentionstat/Glavanostat Autolab (Metrohm) model, PSTAT 30.

The kinetic experiments of the ClER were carried out in pure chloride solutions and obtain a value of the ClER rate costant, the concentration of chloride ions were ranged between (0.5-5M) and changes in the pH values were performed by addition of HCl and maintained between 2-4, because of in the limited range, the ClER is the main electrode reaction. The rest potential of the DSA in solution of (H2O + NaCl+ HCl) as electrolyted was measured against a saturated calomel electrode (SCE) and was varied out by sweeping potentioal between (0.5-1.2 volt). The activation thermodynamic parameters were determined from the ClER rates measured at temperature range of (30-90°C).

The mechanism of the ClER on the DSA is complicated by the specific features of the surface [3], but on the basis of the experminatal result, we tried to propose the mechanism of the ClER on the DSA.

Reference:

1. J. Electrochimica Acta 45 (2000) 2377-2385.
2. J. Braz. Chem. Soc., Vol. 14, No. 3, 388-395, 2003.
3. J. Electrochimica Acta 44 (1999) 2463-2472.

174a Aqueous Partial Molar Volumes from Simulation and Individual Group Contributions

Ashish V. Sangwai, Department of Chemical and Biomolecular Engineering, Tulane University, 300 Lindy Boggs Center, New Orleans, LA 70118 and Hank Ashbaugh, Department of Chemical and Biomolecular Engineering, Tulane University, 6823 St. Charles Avenue, New Orleans, LA 70118.

Partial molar volumes for a range of organic solutes in aqueous solution are evaluated from molecular simulations using Kirkwood-Buff theory. Long-range oscillatory variations in the Kirkwood-Buff integrals are suppressed using the techniques of Lockwood, Rossky, and Levy, enhancing convergence of the calculated partial molar volumes to within the first hydration shell (Daren M. Lockwood and Peter J. Rossky, J. Phys. Chem. B 1999, 103, 1982-1990). Contributions from individual constituent groups, such as methylene carbons and alcohol oxygens, are evaluated using proximal correlation functions extended here to heterogeneous molecular species. We find the simulation partial molar volumes are systematically greater than experiment, with the difference between simulation and experiment increasing with solute size. These differences are attributed to possible deficiencies in the volumetric properties of the interaction potentials employed. Nevertheless, we find that the structure of water near identical groups on different solutes is effectively indistinguishable, as quantified by proximal correlation functions. As a result, we propose a new group contribution correlation for the partial molar volume rooted in Kirkwood-Buff theory and the proximity approximation.

174b Hydrophobic Mismatch Can Induce Specific and Selective Membrane Protein Interactions

Frédérick De Meyer[1], Maddalena Venturoli[1], and Berend Smit[2]. (1) Centre Européen de Calcul Atomique et Moléculaire (CECAM), Ecole Normale Supérieure (ENS), Allée d'Italie 46, Lyon, France, (2) Deptartment of Chemical Engineering, University of California at Berkeley, 20, Gilman Hall, Berkeley, CA 94720-1462

The composition of the lipid membrane and the hydrophobic matching between the lipid bilayer hydrophobic thickness and the hydrophobic length of the transmembrane proteins are important physical properties that regulate the mechanism of lipid-protein interactions in biomembranes. These could play a far more important role in the decision of the protein topology after insertion in the membrane, in the regulation of the protein activity, in protein aggregation of misfolded proteins and in signal transduction, than is presently supposed. However, experimentally it is difficult to isolate and hence to study solely the interactions due to hydrophobic forces. We use free energy calculations and MD simulations of a coarse grain model of a lipid bilayer with embedded proteins to show that hydrophobic forces induce specific and selective protein-protein interactions. The nature of the protein interactions, i.e., repulsion, temporary oligomerization and protein clustering characterized by specific protein cluster sizes, depends on the degree of hydrophobic mismatch, on the geometry of the proteins and on the type of lipid bilayer. We observe that proteins with the same degree of mismatch attract each other selectively until a cluster of a specific size is formed and that protein clusters and proteins with a different degree of mismatch repulse each other. In our approach, we demonstrate how simple hydrophobic interactions can lead to a complex collective behaviour. Our results could improve the understanding of several membrane processes in which protein interactions are involved, and they could have important consequences for future drug design.

174c Multiscale Simulations of Polymer Permeation

Nico F.A. Van der Vegt[1], Vagelis A. Harmandaris[2], Berk Hess[1], Kurt Kremer[1], Tugba A. Ozal[1], and Christine Peter[1]. (1) Max-Planck Institute for Polymer Research, Ackermannweg 10, Mainz, Germany, (2) Max Planck Institute for Polymer Research, Ackermannweg 10, Mainz, Germany

Dynamics and structure formation in complex polymeric systems involves a large range of time and length scales. We discuss multiscale simulation approaches that attempt to link these scales and discuss some of our recent coarse and fine-graining methods applied to polymers in the bulk and polymers in contact with solid substrates. Large-size, fully equilibrated all-atom systems, obtained by fine-graining coarse-grained polymer trajectories, have been employed to study diffusion and solubility of dissolved solute molecules. We discuss how application of Jarzynski's relation (which relates free energies to nonequilibrium work distributions) offers a particularly useful route for calculating free energies of solute molecules dissolved in complex polymer microstructures.

174d Implicit Water Model of Ionic Surfactant Micelles

Arben Jusufi, Antti-Pekka Hynninen, and Athanassios Z. Panagiotopoulos. Chemical Engineering, Princeton University, Engineering Quadrangle, Princeton, NJ 08544

We propose an implicit-solvent model for the simulation of self-assembly of alkali metal dodecyl sulfate micelles. Association of amphiphilic molecules entails multiple time- and length scales. Spontaneous self-assembly of ionic surfactants at low concentrations is currently computationally intractable for atomistic models with explicit solvent. In this work, atomistic Molecular Dynamics (MD) simulations of headgroup and Na counterions are

performed in order to coarse grain the headgroup ion and to trace out the solvent degrees of freedom.

The water-implicit ion interaction potential takes explicitly hydration barriers into account.

For the parameterization of the implicit model of the surfactant tails, we use Grand Canonical Monte Carlo simulations with histogram reweighting techniques.

The matching objectives for the coarse-graining procedure are agreement between simulation and experimental results with regard to their thermodynamic and structural quantities of NaDS micelles. In particular, we seek to reproduce the critical micelle concentration (cmc) and the aggregation number of the resulting micelles. Using this methodology,

we investigate the counterion impact on the micellization process by exchanging Na by Li and K ions. The cmc, aggregation numbers and the degree of counterion binding were obtained for all systems (Li-, Na-, and KDS). The results are in good agreement with experimental results and with recent atomistic MD simulations at high surfactant loadings. The computational methodology can be extended to

other ionic surfactants, including lipids and provides an important step towards large-scale simulations for the study of self-assembly of surfactant molecules.

174e Molecular Thermodynamics of Branched Block Copolymer Micelles

Ramanathan Nagarajan, Molecular Sciences and Engineering Team, Natick Soldier Research, Development & Engineering Center, Kansas Street, Natick, MA 01760

Amphiphilic block copolymer molecules have been shown to form large cylindrical micelles, referred to as wormlike micelles. Under certain conditions, the wormlike micelles transform into branched micelles with the degree of branching increasing from a system where the branched micelles have individual kinetic identities to one in which the branched micelles form an interconnected network. We develop a quantitative treatment of branched micelles formation to establish the conditions under which branching may be initiated and to identify the molecular parameters that determine the extent of branching and network formation in the micellar system. We further extend the treatment to incorporate hydrocarbon solubilizate molecules as a way to control the conditions of formation of branches and micelle networks.

Web Page: www.personal.psu.edu/rxn

174f A Performance Comparison of Equations of State

Ken G. Pepper, Process Engineering, Chicago Bridge and Iron, 350 Pine Street, Suite 400, Beaumont, TX 77701 and R. Tadmor, Chemical Engineering, Lamar University, P.O. Box 10053, Beaumont, Texas 77710.

At the outset of simulating a chemical or petroleum process, the process engineer must select between different property methods. The question that arises is: "Which property method is best for the simulation of this process?" In this article we review past publications regarding the comparison of accuracy of equations of state. We summarize what has been presented to date regarding accuracy of the most commonly used equations of state in process simulation. In addition, we calculate the predicted molar volume of a wide range of pure chemical compounds using four commonly used equations of state: Soave-Redlich-Kwong (SRK), Peng-Robinson (PR), Peng-Robinson-Stryjek-Vera (PRSV), and Lee-Kesler-Plocker (LKP). We compare these predictions with experimental data for these componds and present the results in a convenient tabular form. The purpose of the tabulated results is to assist process engineers in the selection of property methods by providing additional comparative data that has not been provided to date.

174g Cavity Thermodynamics in Organic Solvents

Amit S. Jain, Department of Chemical and Biomolecular Engineering, Tulane University, 300 Lindy Boggs Center, 6823, St. Charles Avenue, New Orleans, LA 70118 and Hank Ashbaugh, Chemical and Biomolecular Engineering, Tulane University, 300 Lindy Boggs Center, 6823, St. Charles Avenue, New Orleans, LA 70118.

Solution thermodynamics is dominated in large part by the process of creating a solute sized cavity in the solvent. Scaled-Particle Theory (SPT) presents fundamental relationships between the thermodynamics of cavity formation and spontaneous solvent fluctuations. We have studied using Molecular Dynamics simulations a number of organic solvents the statistics of cavity formation. These simulation results are then compared to a revised version of Reiss's classic SPT that takes into account the molecular shape of the solvent. Carbon atom densities in the vicinity of the cavity surface, i.e., the contact density, have been evaluated for these solvents. Systematic changes in the contact density for homologous aliphatic hydrocarbons observed from simulation are correctly reproduced from our molecular SPT. Moreover, the molecular SPT accurately predicts the ordering of contact densities for polyethylene oligomer melts of increasing length, that would not be captured by Reiss's classic SPT which ascribes differences in solvents simply to differences in an empirical van der Waals solvent radius. Our molecular theory, on the other hand requires only one van der Waals radius for the methyl/ methylene carbons and is then predictive for a wide range of solvents without resorting to post hoc empirical parameterizations.

174h Prediction and Modeling of Solubility Phenomena in Subcritical Fluids Using An Extended Solubility Parameter Approach

Keerthi Srinivas[1], Jerry W. King[1], and Charles M. Hansen[2]. (1) Chemical Engineering, University of Arkansas, 3202 Bell Engineering, Fayettville, AR 72701, (2) Charles M. Hansen Consulting, Jens Bornøsvej 16, 2970, Hørsholm, Denmark

There has been a growing interest in use of sub-critical fluids in extraction and/or reactive processing of natural products. Subcritical water and hydroethanolic solvents under compression above their boiling points are currently being used both in the extraction of pharmacological/nutraceutical components as well as for biomass/bioenergy conversion applications. The first step towards design and optimization of such systems is the assessment of physicochemical properties of solutes in sub-critical solvents. In the past, we have previously correlated solute solubility in subcritical fluids using different solubility models like those of Clifford-Hawthorne[1] and delValle et. al.[2]. As an extension of these studies, we have investigated the applicability of modified solubility parameter approaches including the Hansen 3-dimensional solubility parameters and the RED theory. As to be expected, the dispersion, polar and hydrogen bonding partial solubility parameter components of subcritical solvents decrease with temperature to varying degrees. There exists insufficient experimental data for the partial solubility parameters of certain complex structural solutes in the literature. Hence, group contribution methods and suitable equations based on molecular properties were used to calculate the partial solubility parameters for such solutes. Relative Energy Difference (RED) numbers reflects the affinity of the solutes for subcritical fluids depending on their values at various temperatures. A SPHERE[3] software program can be used to generate three dimensional solubility parameter spheroid maps that can help in studying the interactions between the solutes and the subcritical solvent. The above approach when coupled with a group contribution method and/or a statistical thermodynamic-based or corresponding states approach can be used to estimate solubility parameter trends in subcritical fluid media. These trends coupled with the study of the temperature and pressure dependence on the partial solubility parameters of structurally-complex solutes and pressurized/subcritical solvents can be used for modeling extraction or reaction selectivity of solutes in subcritical solvents.

[1] Anthony A. Clifford and Steven B. Hawthorne, "Processes in Subcritical Water", Super Green 2002, Proceedings of the 1st International Symposium on Supercritical Fluid Technology for Energy and Environment Applications, Kyung Hee University, Suwon, Korea, November 3-6,2002, pp. 21-29.

[2] Jose M del Valle, Juan C. de la Fuente and Jerry W. King, "Correlation for the Variations with Temperatura of Solute Solubilities in Hot, Pressurized Water", Equiphase Conference, Morelia, Michoacán, México, October 21-25, 2006 (To be published).

[3] Charles M. Hansen, "Hansen Solubility Parameters: A User's Handbook", CRC Press, Boca Raton, FL, 1999.

211a Thermodynamic Testing Methods in Energetic Material Evaluation

Matthew Hafner, Randy Busky, and Mark Mansfield. Technology, ATK, LCAAP, HWY 7 & 78, Independence, MO 64051

The ATK technology group at Lake City Army Ammunition Plant (LCAAP) has developed a new method to incorporate thermal property testing into new material qualification testing. Current LCAAP raw material testing has resulted in materials passing specification testing, yet failing ballistic functional testing. LCCAP conducts testing according to military specifications on candidate materials prior to a ballistics work test. This old test and shoot process analyzes materials for chemical make up and in some cases particle size prior to the functionality testing. With the addition of thermal testing, we can improve the reliability of our new material screening process.

LCAAP mixes solid raw materials into energetic mixes that very rapidly change phase from all solid reactants to gas, liquid and solid products in order to function ammunition properly. For these energetic materials, we need to ensure that we have the right amount of energy available in the system delivered at the proper rate for desired function. We plan to use different thermodynamic techniques to estimate the Gibbs free energy, enthalpy, and entropy. Bomb Calorimetry will provide an estimate of the total energy available in the system, DSC and TGA will provide us with estimate on the enthalpy and entropy. We need these estimates to determine if the required energy is available and if the entropy and enthalpy are in the proper ratio to allow the desired rate transfer to occur. These thermodynamic tests also give us insight into the sensitivity of the new materials and the resulting mixes. Contaminants can change: 1) the onset temperature of mixes, 2) the sensitivity to different initiation mechanisms (i.e. electrostatic discharge, friction and impact), 3) the rate of output and 4) the amount of energy output. Understanding and predicting these responses helps to ensure safe operation of our current processes and aid new mix development.

We may have a situation where we cannot find a match for the mil-spec material. Using the thermodynamic fingerprinting of the baseline material, we will be able to propose methods to "clean up" contaminated sources or seed purer sources with the required chemicals and prove out the thermal response of the new material to meet the mil-spec. Most of our materials are refined from various natural sources and have differing levels of contamination. Using L. Bruce Railsback's "An Earth Scientist's Periodic Table of the Elements and Their Ions" we can use the responses on the DSC and TGA scans to identify potential contaminates that may not be found in the current specification testing1.

The small quantities required for these tests will allow us to test gram quantities of materials in a shorter timeframe for screening instead of relying on the pound quantities required for ballistics testing. This will help to speed up and improve new product development, and increase our understanding of the physics behind the materials' function. We will still use ballistics testing as a final prove out of the new materials, but we will have increased our probability of success and limited the resources necessary to screen new materials by using these thermal tests.

1) Railsback, L. Bruce; "An Earth Scientist's Periodic Table of the Elements and Their Ions", University of Georgia, Athens, GA, 2004.

211b Models for Infinite Dilution Activity Coefficients in Ternary Liquid Systems

akand W. Islam, Department of Chemical Engineering, North Carolina A&T State University, 421 E Hendrix, Apt#D, Greensboro, NC 27405 and Vinayak N. Kabadi, Chemical Engineering Department, North Carolina A&T State University, McNair Hall, Greensboro, NC 27411.

The objective of this research was to study the change in partition coefficient with concentration of solute in a liquid system consisting of two relatively immiscible solvents. This would require a consistency between data for infinite dilution partition coefficients and ternary LLE data at finite concentrations of the solute. Moreover an attempt would be made to model the infinite dilution and finite concentration behavior of these systems by a single thermodynamic model. During the course of our work, we found 35 ternary systems in the literature for which the infinite dilution partition coefficients and the ternary LLE data could not be predicted accurately by the currently available liquid state models.

Ternary liquid-liquid equilibrium data for hexane+1-butanol+water, carbon tetrachloride +propanoic acid +water, and water+propanoic acid+hexane were measured from very low concentrations of the solute to finite concentrations at 25 C. The measured partition coefficients showed good agreement with the literature values. UNIQUAC and modified UNIQUAC activity coefficient models were used to represent the data. The results showed large errors, exhibiting the inability of these models to correlate the data. Attempts are being made to apply some modified versions of quasi-chemical activity coefficient models to these systems.

211c Modeling the Dynamic Viscosity of Ionic Liquids

Aimee Ruiz-Llamas and **Ricardo Macías-Salinas**. ESIQIE-Departamento de Ingeniería Química, Instituto Politécnico Nacional, UPALM, Av. IPN s/n, Col. Lindavista, México, D.F., 07738, Mexico

A newly developed thermodynamic model for the accurate correlation and/or prediction of the dynamic viscosity of ionic liquids is discussed here. The viscosity model is based on the Eyring's absolute rate theory for pure liquids where the activation energy for viscous flow was related to well-known thermodynamic quantities readily obtained from simple cubic equations of state of the van der Waals type. The resulting model was successfully validated during the representation of experimental dynamic viscosities of various imidazolium- and pyridinium-based ionic liquids within a wide temperature range and at atmospheric pressure. Model viscosity extrapolations at higher pressures were also verified.

211d Predicting Phase Equilibrium near the Critical Point Using the Soave-Redlich-Kwong Equation of State

Johnelle M. Halteman, Chemistry and Physics, Saint Mary's College, Notre Dame, IN 46556

The objective of this study was to improve methods for predicting vapor pressure from the Soave-Redlich-Kwong (SRK) cubic equation of state (EOS) at temperatures near the critical point. The critical point of a substance is the specific temperature and pressure above which vapor and liquid phases are no longer distinguishable. A method for vapor pressure prediction from the SRK equation at low temperatures was adapted and combined with existing results near the critical point to derive a new vapor pressure function. By relating two different Taylor series expansions for the natural logarithm of the reduced pressure, the coefficients of the vapor pressure function and their dependence on the acentric factor of the substance were determined. Vapor pressure predictions were within a factor of two for reduced temperatures above 0.54 when the acentric factor was 0.8 and for reduced temperatures above 0.41 when the acentric factor was zero. This was a significant improvement on existing methods, based upon Taylor series expansions.

211e A Relative Volatility Scale for Ionic Liquids

Joseph W. Magee[1], Jason A. Widegren[1], Wesley A. Henderson[2], and Yi-Ming Wang[3]. (1) Physical & Chemical Properties Division, National Institute of Standards and Technology, Mail Stop 838.07, 325 Broadway, Boulder, CO 80305-3328, (2) Department of Chemistry, U.S. Naval Academy, 572 M Holloway Road, Stop 9B, Annapolis, MD 21402, (3) Department of Chemistry and Chemical Biology, Harvard University, Cambridge, MA 02138
An earlier collaborative work [1] demonstrated laboratory distillation of ionic liquids and established the feasibility of separative distillation of ionic liquid mixtures. As a continuation of the earlier work, relative volatilities of a variety of common, aprotic ionic liquids were determined for the first time [2]. Equimolar mixtures of ionic liquids were vacuum distilled in a glass sublimation apparatus at approximately 473 K. The composition of the initial distillate, determined by NMR spectroscopy, was used to establish the relative volatility of each ionic liquid in the mixture. The effect of alkyl chain length was studied by distilling mixtures of 1-alkyl-3-methylimidazolium bis(trifluoromethylsulfonyl)imide ionic liquids, or mixtures of N-alkyl-N-methylpyrrolidinium bis(trifluoromethylsulfonyl)imide ionic liquids, shown in Fig. 1, with different alkyl chain lengths. For both classes of salts the volatility is highest when the alkyl side chain is a butyl group. The effect of cation structure on volatility has been determined by distilling mixtures containing different types of cations. Generally speaking, ionic liquids based on imidazolium and pyridinium cations are more volatile than ionic liquids based on ammonium and pyrrolidinium cations, regardless of the types of counter ions present. Similarly, ionic liquids based on the anions [(C2F5SO2)2N]-, [(C4F9SO2)(CF3SO2)N]- and [(CF3SO2)2N]-; are more volatile than ionic liquids based on [(CF3SO2)3C]- and [CF3SO3]-, and are much more volatile than ionic liquids based on [PF6]-. The laboratory demonstration of RTIL volatility has opened a new branch of scientific research on ionic vapors and unlocked doors leading to potential applications, including new routes to purify ionic liquids to unprecedented levels, and to regenerate spent ionic liquids from industrial processes. References [1] Earle, M. J., Esperança, J. M. S. S., Gilea, M. A., Canongia Lopes, J. N., Rebelo, L. P. N., Magee, J. W., Seddon, K. R., and Widegren, J A., "The Distillation and Volatility of Ionic Liquids" Nature 439: 831-834 (2006). [2] Widegren, J. A., Wang, Y.-M., Henderson, W. A., and Magee, J. W., "Relative Volatilities of Ionic Liquids by Vacuum Distillation of Mixtures" J. Phys. Chem. B 111: 8959-8964 (2007).

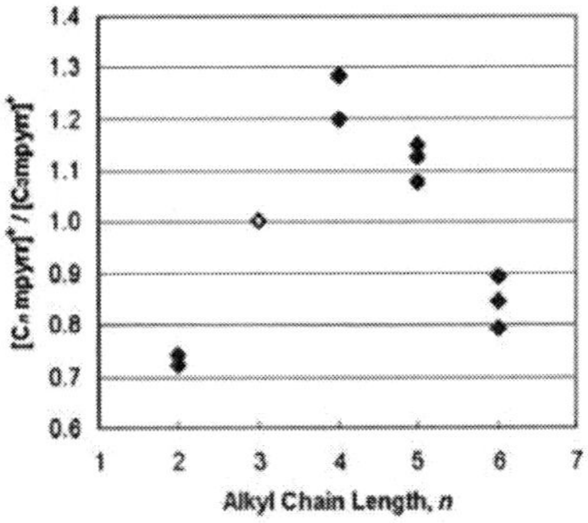

Fig. 1.

211f Estimation of the Isothermal Compressibility and the Thermal Expansion Coefficient Using Density Explicit Empirical Equations

Abel Zúñiga-Moreno, Departamento de Ingeniería Química, Escuela Superior de Ingeniería Química e Industrias Extractivas, Instituto Politecnico Nacional, Edif 7, 1er Piso, UPALM, Lindavista, Mexico, 07738, Mexico

The isothermal compressibility and thermal expansion coefficient for diverse substances were estimated using explicit volume empirical equations. Even though most empirical equations can represent densities appropriately [1,2], their use to calculate 2nd order thermodynamic properties has not been studied deeply. In this work, Tait-type equations and short explicit volume equations were tested to know their capabilities to estimate the isothermal compressibility and thermal expansion coefficient. The empirical equations were used to solve the density derivatives respect to pressure and respect to temperature in order to obtain isothermal compressibility and thermal expansion coefficient. Estimated values using the empirical equations were compared with values obtaining using the pρT raw data.

211g Correlation of Liquid Thermophysical Properties Via Virial-Based Mixing Rules

Srividya Gummadi and Sabyasachi Sen. Invensys SimSci-Esscor, 26561 Rancho Parkway South, Suite 100, Lake Forest, CA 92630

Over the last several decades, a large number of methods have been proposed for calculating the thermophysical properties of pure liquids (and gases). These include corresponding states techniques and statistical thermodynamics approaches. In modern simulators these methods are usually referred to as stand-alone or library methods (i.e., not based on multi-property P-V-T equations of state), and are available as a standard set of "temperature-dependent correlations". The component-specific parameters required by these correlations are implemented as built-in or user-defined libraries. A typical example of a built-in library is one that is derived from the DIPPR-801 database - a collection of high-quality peer-reviewed component-specific parameters.

Although, these library correlations provide high quality estimates for several pure component liquid properties of interest, the science for describing mixtures has consistently trailed behind. For example, the mixing rules typically implemented in process simulators for liquid viscosity and thermal conductivity, are the Kendall-Monroe (1917) and Li (1976) correlations, respectively. While the attractive feature of these mixing rules is that they do not require interaction parameters (IP's), their downside is that the do not allow IP's, even when necessary. In many applications, they incorrectly predict only a small deviation from ideal mixing, often with incorrect sign. Thus, while these mixing rules perform reasonably for hydrocarbon systems, their efficacy for non-ideal systems is questionable.

The state-of-the-art is the Redlich-Kister correlation (RKC), which can incorporate additional higher-order terms to improve the fit of data with higher precision. The RKC is often the correlation of choice of the applied thermodynamics research community. It has been successfully applied to correlate a variety of thermophysical properties for binary mixtures, including density, enthalpy, and viscosity. However, the RKC does not naturally extend to multi-component systems. In addition, cross-correlation versus temperature compromises its ability to describe composition-dependence.

This investigation draws upon the exact science of mixing rules that has been established for the virial coefficients of gases. Perturbations from ideal mixing are modeled akin to successive terms in the virial expansion, resulting in the name "Virial-Based Mixing Rules" (VBMR). Each perturbation reduces to zero for pure components, and adopts the composition order of the corresponding virial coefficient it represents. For example, the first perturbation represents the second virial coefficient and implements a quadratic mixing rule, while allowing one adjustable binary interaction parameter. The next term employs cubic mixing and allows multiple interaction parameters, including a ternary interaction parameter which may be fine-tuned for ternary and higher mixtures. All interaction parameters are initialized to zero.

Several binary and ternary systems (featuring high-precision as well as noisy data) are investigated. The target properties in this study are liquid density, enthalpy, viscosity and thermal conductivity. The necessity of the higher-order terms in correlating precise data within measurement uncertainty is established. In contrast, it is demonstrated how data with excessive noise do not warrant correlation with higher-order terms, while noting that the proposed approach has a low susceptibility to correlate noise. Wherever appropriate, the VBMR are compared with the RKC (for example, in the correlation of precise data). Cross-correlation with temperature resulting in a compact set of parameters is also demonstrated.

211h A New Method for Calculating Gas Minimum Miscibility Pressure

Adel Elsharkawy, Petroleum Engineering Department, Kuwait University, B.O. Box 5969, Safat 13060, Kuwait

Minimum miscibility pressure is a key parameter for the design and operations of successful gas miscible flood project to enhance oil recovery. Operating below the minimum miscibility pressure (MMP) results in immiscible displacement, gas fingering and early breakthrough, and consequently low oil recovery. Operating above the MMP increases project expenses with little or no additional oil recovery.

This paper presents a new method for calculating the MMP taken into consideration the effect of various components of volatile components (C1-N2), intermediate components (C2-C6, H2S), and heavy (C7+). Thus the new method takes into consideration the effect of gas-oil ratio on the MMP.

Accuracy of the proposed method has been compared to statistical correlations and equation of state based methods. The comparison indicates that the new method successfully and accurately predict MMP better than the other published methods. The new methods can be used to estimate MMP for planning miscibility tests, predict miscibility pressure when reliable experimental data is unavailable, and screening oil reservoirs for potential enhanced oil recovery project for specific gas injection.

AUTHOR INDEX

American Institute of Chemical Engineers
3 Park Avenue
New York, New York 10016-5991

ISBN: 978-1-60560-217-2